Coliforms in Distribution Systems: Integrated Disinfection and Antimicrobial Resistance

Awwa
Research
Foundation

About the Awwa Research Foundation

The Awwa Research Foundation (AwwaRF) is a member-supported, international, nonprofit organization that sponsors research to enable water utilities, public health agencies, and other professionals to provide safe and affordable drinking water to consumers.

The Foundation's mission is to advance the science of water to improve the quality of life. To achieve this mission, the Foundation sponsors studies on all aspects of drinking water, including supply and resources, treatment, monitoring and analysis, distribution, management, and health effects. Funding for research is provided primarily by subscription payments from approximately 1,000 utilities, consulting firms, and manufacturers in North America and abroad. Additional funding comes from collaborative partnerships with other national and international organizations, allowing for resources to be leveraged, expertise to be shared, and broad-based knowledge to be developed and disseminated. Government funding serves as a third source of research dollars.

From its headquarters in Denver, Colorado, the Foundation's staff directs and supports the efforts of more than 800 volunteers who serve on the board of trustees and various committees. These volunteers represent many facets of the water industry, and contribute their expertise to select and monitor research studies that benefit the entire drinking water community.

The results of research are disseminated through a number of channels, including reports, the Web site, conferences, and periodicals.

For subscribers, the Foundation serves as a cooperative program in which water suppliers unite to pool their resources. By applying Foundation research findings, these water suppliers can save substantial costs and stay on the leading edge of drinking water science and technology. Since its inception, AwwaRF has supplied the water community with more than $300 million in applied research.

More information about the Foundation and how to become a subscriber is available on the Web at **www.awwarf.org.**

Coliforms in Distribution Systems: Integrated Disinfection and Antimicrobial Resistance

Prepared by:
Graham A. Gagnon, **Heather M. Murphy**, **Jennie L. Rand**, and **Sarah Jane Payne**
Dalhousie University
Department of Civil & Resource Engineering
1360 Barrington Street, Halifax, Nova Scotia B3J 2X4 CANADA
and
Susan Springthorpe, **Fernando Matias**, **Reza Nokbeh**, and **Syed A. Sattar**
University of Ottawa
Centre for Research on Environmental Microbiology
451 Smyth Road, Ottawa, Ontario, K1H 8M5 CANADA

Sponsored by:
Awwa Research Foundation
6666 West Quincy Avenue, Denver, CO 80235-3098

Published by:

Awwa Research Foundation

American Water Works Association

IWA Publishing

DISCLAIMER

This study was funded by the Awwa Research Foundation (AwwaRF). AwwaRF assumes no responsibility for the content of the research study reported in this publication or for the opinions or statements of fact expressed in the report. The mention of trade names for commercial products does not represent or imply the approval or endorsement of AwwaRF. This report is presented solely for informational purposes.

CONTENTS

LIST OF TABLES vii

LIST OF FIGURES xi

FOREWORD xv

ACKNOWLEDGMENTS xvii

EXECUTIVE SUMMARY xix

CHAPTER 1: PROJECT OVERVIEW 1
- Introduction 1
- Research Objectives 2
- Report Organization 3

CHAPTER 2: LITERATURE REVIEW 5
- Biofilm Composition 5
- Coliform Occurrence/Regrowth In Distributions Systems 5
- Health Significance Of Coliforms In Distributions Systems 6
- Disinfection Strategies 7
  - Ultraviolet Light 7
  - Chemical Disinfectants 8
- Microbial Stressors And Stress Responses In Drinking Water 8

CHAPTER 3: MATERIAL AND METHODS 11
- Introduction 11
- Phase 1 – Bench Scale Experiments 11
  - Experimental Design 11
  - Bench –Scale Distribution System Characteristics 16
  - Influent Water Characteristics 16
  - *E. coli* Spike Procedure 17
  - Disinfectant Preparation 18
- Phase 2 - Field Scale Experiments 19
  - Experimental Design 19
  - Field Scale Water Quality 21
- Sample Collection And Analysis 22
  - Microbial Enumeration and Characterization 22
  - Biofilm Community Composition 25
  - Statistical Analyses 33

CHAPTER 4: LABORATORY SCALE DISINFECTION STUDIES 35
Introduction 35
Results 35
Disinfectant Residuals 35
Effluent Heterotrophic Bacteria 41
Effect of UV Disinfection 41
Biofilm Heterotrophic Bacteria 47
Effluent *E. coli* 54
Analysis of Suspended *E. coli* 62
Biofilm *E.coli* 67
Analysis of Biofilm *E. coli* 75
Coliform Ecology-Phenotyping 79
DGGE Analysis 85
Acquisition of Antibiotic Resistance in DWDS Biofilms 96
Summary 105

CHAPTER 5 FIELD SCALE DISINFECTION STUDIES – SURFACE WATER 109
Introduction 109
Results and Discussion 109
Acclimation/Pre-disinfection period 109
Post-disinfection HPC data 109
Comparison of Disinfectants 112
Water Quality Analysis 114
Summary 115

CHAPTER 6 FIELD SCALE DISINFECTION STUDIES – GROUNDWATER 117
Introduction 117
Results And Discussion 118
Comparison of Disinfectants 118
Water Quality Analysis 122
Summary 126

CHAPTER 7 PROJECT SUMMARY AND RECOMMENDATIONS 127
Project Summary 127
Recommendation For The Water Industry 129

REFERENCES 131

ABBREVIATIONS 139

TABLES

3.1	Experimental factors……	12
3.2	Target Disinfectant Residuals……	12
3.3	Average Water Quality for the GAC/BAC treated water……	17
3.4	Average Water Quality for the UV treated water……	17
3.5	Average Water Quality for the Raw and UV Treated Water……	22
3.6	Average Water Quality for the Raw Chloraminated and ChlorinatedWater……	22
4.1	Target Disinfectant Residuals……	36
4.2	Disinfectant residuals for chlorine trial……	37
4.3	Disinfectant residuals for chlorine dioxide trial……	38
4.4	Disinfectant residuals for chloramine trial……	39
4.5	Effluent HPC log reductions……	41
4.6	Biofilm HPC log reductions……	48
4.7	Effluent *E.coli* log Reductions……	55
4.8	Changes in effluent HPC concentrations after the *E. coli* spikes……	55
4.9	P-values for the difference between *E. coli K12* and Mutant strain on suspended *E. coli* counts in cast iron and polycarbonate ARs……	57
4.10	P-values for the difference between UV treatments on suspended *E. coli* counts in cast iron and polycarbonate ARs……	62
4.11	P-values for the difference between cast iron and polycarbonate coupons on suspended *E. coli* counts……	63
4.12	Biofilm *E. coli* log Reductions……	67
4.13	Changes in biofilm HPC concentrations after the *E. coli* spikes……	68

4.14 P-values for the difference between *E. coli K12* and the mutant strain on biofilm *E. coli* counts in cast iron and polycarbonate ARs (shaded areas indicate significance)... 70

4.15 P-values for the difference between UV treatments on biofilm *E. coli* counts in cast iron and polycarbonate ARs.. 72

4.16 P-values for the difference between cast iron and polycarbonate coupons on biofilm *E. coli* counts.. 75

4.17 API 20E Substrate identification table 79

4.18 API 20E test kit results for *E. coli K12* during the chlorine trial..................... 81

4.19 API 20E test kit results for *E. coli MMR* mutant during the chlorine trial.......... 81

4.20 API 20E test kit results for *E. coli K12* during the chlorine dioxide trial........... 82

4.21 API 20E test kit results for *E. coli MMR* mutant during the chlorine dioxide trial.. 83

4.22 API 20E test kit results for *E. coli K12* during the chloramines trial................ 84

4.23 API 20E test kit results for *E. coli MMR* mutant during the chloramines trial..... 84

4.24 Identification and Comparison of Bands by PCR-DGGE for cast iron and polycarbonate.. 86

4.25 Identification and Comparison of Bands by PCR-DGGE for the Chloramine Run.. 87

4.26 Identification and Comparison of Bands by PCR-DGGE for the Chlorine Dioxide Run.. 88

4.27 Identification and Comparison of Bands by PCR-DGGE for the Free Chlorine Run.. 89

4.28 The numbers of isolates screened and the percentage that were not *E. coli* from m-Endo plates.. 97

4.29 Multiple antibiotics resistant *E. coli K12* and *E. coli.* muts isolates from untreated control and chlorine treated biofilm reactors............................... 99

4.30 Multiple antibiotics resistant *E. coli K12* and *E. coli.* muts isolates from untreated control and chloramine treated biofilm reactors............................ 100

4.31 Multiple antibiotics resistant *E. coli K12* and *E. coli.* muts isolates from untreated control and chlorine dioxide treated biofilm reactors...................... 101

4.32 Overall relative co-occurrence of Rifampicin-resistant phenotype among multiple resistant isolates.. 102

4.33 Use of cast iron (CI) or polycarbonate (PC) coupons in reactors and the coincidence of resistance to Rifampicin and multiple antibiotics for reactors with upstream UV treatment... 103

4.34 Relative occurrence of rif co-resistant isolates among multiple resistant isolates. 105

5.1 Suspended and Attached HPC Bacteria Pre- and Post-Disinfection................ 110

FIGURES

3.1	Bench Scale setup	13
3.2	Pre-UV treatment setup	14
3.3	Non-UV treatment setup	15
3.4	Post-UV treatment setup	15
3.5	Schematic of the Halifax Field Scale Setup	20
3.6	Schematic of Pinellas County Field Scale Setup	21
3.7	Design for *rpoB* primers	33
4.1	Average disinfectant residuals in cast iron ARs	40
4.2	Average disinfectant residuals in polycarbonate ARs	40
4.3	Suspended heterotrophic bacteria in cast iron ARs during chlorine disinfection	42
4.4	Suspended heterotrophic bacteria in polycarbonate ARs during chlorine disinfection	43
4.5	Suspended heterotrophic bacteria in cast iron ARs during chlorine dioxide disinfection	44
4.6	Suspended heterotrophic bacteria in polycarbonate ARs during chlorine dioxide disinfection	45
4.7	Suspended heterotrophic bacteria in cast iron ARs during chloramine disinfection	46
4.8	Suspended heterotrophic bacteria in polycarbonate ARs during chloramine disinfection	47
4.9	Biofilm heterotrophic bacteria in cast iron ARs during chlorine disinfection	49
4.10	Biofilm heterotrophic bacteria in polycarbonate ARs during chlorine disinfection	50
4.11	Biofilm heterotrophic bacteria in cast iron ARs during chlorine dioxide disinfection	51
4.12	Biofilm heterotrophic bacteria in polycarbonate ARs during chlorine dioxide disinfection	52

4.13 Biofilm heterotrophic bacteria in cast iron ARs during chloramine disinfection. 53

4.14 Biofilm heterotrophic bacteria in polycarbonate ARs during chloramine disinfection.. 54

4.15 Effluent HPCs for cast iron ARs spiked with *E. coli K12* during the chlorine trial (arrows indicate when spikes occurred)... 56

4.16 Effluent HPCs for polycarbonate ARs spiked with *E. coli K12* during the chlorine trial.. 56

4.17 Suspended *E. coli K12* in cast iron ARs during chlorine disinfection.............. 58

4.18 Suspended *E. coli K12* in polycarbonate ARs during chlorine disinfection........ 58

4.19 Suspended *E. coli K12* in cast iron ARs during chlorine dioxide disinfection..... 59

4.20 Suspended *E. coli K12* in polycarbonate ARs during chlorine dioxide disinfection... 60

4.21 Suspended *E. coli K12* in cast iron ARs during chloramine disinfection........... 61

4.22 Suspended *E. coli K12* in polycarbonate ARs during chloramine disinfection.... 61

4.23 Suspended *E. coli K12* in cast iron ARs.. 65

4.24 Suspended *E. coli K12* in polycarbonate ARs.. 66

4.25 Biofilm HPCs for cast iron ARs spiked with *E. coli K12* during the chlorine trial (arrows indicate when spikes occurred)... 68

4.26 Biofilm HPCs for polycarbonate ARs spiked with *E. coli K12* during the chlorine trial (arrows indicate when spikes occurred)............................... 69

4.27 Biofilm *E. coli K12* in cast iron ARs during chlorine disinfection................. 71

4.28 Biofilm *E. coli K12* in polycarbonate ARs during chlorine disinfection........... 71

4.29 Biofilm *E. coli K12* in cast iron ARs during chlorine dioxide disinfection........ 72

4.30 Biofilm *E. coli K12* in polycarbonate ARs during chlorine dioxide disinfection.. 73

4.31 Biofilm *E. coli K12* in cast iron ARs during chloramines disinfection............. 74

4.32 Biofilm *E. coli K12* in polycarbonate ARs during chloramines disinfection....... 74

4.33 Biofilm *E. coli K12* in cast iron ARs... 77

4.34 Biofilm *E. coli K12* in polycarbonate ARs.. 78

4.35 DGGE gel of PCR products from DNA extracted from the biofilm - pre-treatment phase of the 2nd. Chlorine experiment................................... 90

4.36 DGGE gel of PCR products from DNA extracted from the biofilm – low chlorine phase of the 2nd. Chlorine experiment....................................... 91

4.37 DGGE gel of PCR products from DNA extracted from the biofilm – pretreatment phase of chloramine experiment.. 92

4.38 DGGE gel of PCR products from DNA extracted from the biofilm – low chloramine phase of the chloramine experiment..................................... 93

4.39 DGGE gel of PCR products from DNA extracted from the biofilm – low chloramine phase of the chloramine experiment..................................... 94

4.40 DGGE gel of PCR products from DNA extracted from the biofilm – low chlorine dioxide phase of the chlorine dioxide experiment.......................... 95

4.41 PCR products generated from cluster II of rpoB gene (A); BclI digestion of PCR products (B) 105

5.1 Pockwock Suspended Heterotrophic Bacteria Counts............................ 111

5.2 Pockwock Attached Heterotrophic Bacteria Counts............................... 112

5.3 Pockwock Suspended Heterotrophic Bacteria Counts............................ 113

5.4 Pockwock Attached Heterotrophic Bacteria Counts............................... 114

6.1 Average Suspended Heterotrophic Bacteria in the Influent Streams............... 119

6.2 Free Chlorine Residual with Corresponding HPC Bacteria in the Influent of the AR Receiving Water Treated with UV and Cl_2...................................... 120

6.3 Free Chlorine Residual with Corresponding HPC Bacteria in the Influent of the AR Receiving Water Treated with Cl_2 Only.. 120

6.4 Average Suspended Heterotrophic Bacteria in the Effluent of Each AR........... 121

6.5 Average Attached Heterotrophic Bacteria from PVC Coupons..................... 122

6.6 Ammonia, Nitrite and Nitrate Concentrations in the Effluent of the AR Treated with NH_2Cl Only.. 125

6.7 Ammonia, Nitrite and Nitrate Concentrations in the Effluent of the AR Treated with UV and NH_2Cl.. 126

FOREWORD

The Awwa Research Foundation is a nonprofit corporation that is dedicated to the implementation of a research effort to help utilities respond to regulatory requirements and traditional high-priority concerns of the industry. The research agenda is developed through a process of consultation with subscribers and drinking water professionals. Under the umbrella of a Strategic Research Plan, the Research Advisory Council prioritizes the suggested projects based upon current and future needs, applicability, and past work; the recommendations are forwarded to the Board of Trustees for final selection. The foundation also sponsors research projects through the unsolicited proposal process; the Collaborative Research, Research Applications, and Tailored Collaboration programs; and various joint research efforts with organizations such as the U.S. Environmental Protection Agency, the U.S. Bureau of Reclamation, and the Association of California Water Agencies.

This publication is a result of one of these sponsored studies, and it is hoped that its findings will be applied in communities throughout the world. The following report serves not only as a means of communicating the results of the water industry's centralized research program but also as a tool to enlist the further support of the nonmember utilities and individuals.

Projects are managed closely from their inception to the final report by the foundation's staff and large cadre of volunteers who willingly contribute their time and expertise. The foundation serves a planning and management function and awards contracts to other institutions such as water utilities, universities, and engineering firms. The funding for this research effort comes primarily from the Subscription Program, through which water utilities subscribe to the research program and make an annual payment proportionate to the volume of water they deliver and consultants and manufacturers subscribe based on their annual billings. The program offers a cost-effective and fair method for funding research in the public interest.

A broad spectrum of water supply issues is addressed by the foundation's research agenda: resources, treatment and operations, distribution and storage, water quality and analysis, toxicology, economics, and management. The ultimate purpose of the coordinated effort is to assist water suppliers to provide the highest possible quality of water economically and reliably. The true benefits are realized when the results are implemented at the utility level. The foundation's trustees are pleased to offer this publication as a contribution toward that end.

Walter J. Bishop
Chair of Board of Trustees
Awwa Research Foundation

Robert C. Renner, P.E.
Executive Director
Awwa Research Foundation

ACKNOWLEDGMENTS

The members of the project team are grateful for the financial support provided by the Awwa Research Foundation. In particular, we acknowledge the efforts of our project manager, Dr. My-Linh Nguyen. As well, we also acknowledge the insightful technical contributions made by our project advisory committee.

The project team is also thankful to Trojan Technologies Ltd. for providing a bench-scale UV reactor and invaluable technical support to conduct the research. Finally, the team is very appreciative of the in-kind support provided by our utility partners, namely Pinellas County Utilities and Halifax Regional Water Commission. In particular, we are very grateful for the sampling and diligent operations provided by Marsha Pryor and Donna Mooren in Pinellas County and Peter Flinn in Halifax.

EXECUTIVE SUMMARY

This study aimed to significantly expand the current database between UV and chemical disinfectants by examining their combined effects on *Escherichia coli* (*E. coli*) spiked into mixed species biofilm, and by scrutinizing any observed synergistic effects through the use of advanced molecular techniques. In particular we examined the antibiotic resistance of any coliforms surviving low concentrations of disinfectant and also the composition of the naturally-formed biofilms. Most studies on distribution system biofilms have used only culture to determine bacteria present. More recently developed molecular techniques (e.g. PCR-DGGE) offer some advantages in permitting determination of biofilm composition and dominant organisms, even if these cannot be isolated in culture (Muyzer, 1999). Schwartz et al. (2003) used a combination of culture and molecular techniques to follow changes in biofilm composition with UV and ClO_2 application, recording increased metabolic activity in biofilms treated with UV, and that ClO_2 was critical for mitigating pathogens such as enterococci although neither disinfection method could deal with more resistant pathogens such as legionellae and mycobacteria. No comparison of ClO_2 to Cl_2 and chloramines was conducted in that study. Pryor et al (2003) used a combination of culture and molecular techniques to follow changes in biofilm composition with changes in disinfectant residual (chlorine and chloramine) and water source, this included some important changes in the opportunistic pathogens legionellae and mycobacteria. No UV disinfection was included in this study. Molecular techniques have also been used to follow composition and succession in biofilms in a model drinking water biofilm from 1 day to 3 years (Martiny et al., 2003). In addition a recent AwwaRF project (Gagnon et al., 2004) found that UV disinfection in combination with ClO_2 or Cl_2 resulted in an observed synergy, as quantified exclusively by heterotrophic bacteria. Synergistic effects between UV/ClO_2 and UV/Cl_2 resulted in greater disinfection in the distribution system than a single disinfectant (i.e., UV, ClO_2, Cl_2). It is hypothesized that molecular techniques will help to explain the synergies observed previously, particularly as they relate to potential pathogens.

Research Objectives

- Provide an overview of biofilm formation and composition, coliform occurrence and associated health significance, various disinfection strategies, and insight into the microbial stress response to stressors in drinking water.
- Provide a detailed explanation of the materials and methods used throughout the experimental study. Design and operational parameters for both the bench-scale and field scale experiments are detailed, as well as relevant information about the system characteristics.
- Evaluate the impact of the various disinfection schemes and pipe material on both heterotrophic plate bacteria and *E.coli* at the bench scale. The phenotypic response and antimicrobial resistance of *E.coli* is also evaluated at the bench scale. Evaluate the field scale results are compare the affects of the different disinfectants on heterotrophic bacteria.
- Evaluate the impact of disinfectant residual and upstream UV treatment on both effluent and biofilm *E. coli* in field scale study from a surface water source.

- Evaluate the impact of disinfectant residual and upstream UV treatment on both effluent and biofilm *E. coli* in field scale study groundwater source in a warm climate.
- Provide the overall project summary and recommendations to the water industry in terms of practical implications of the research.

APPROACH

Laboratory and field studies were performed in parallel to model pathogens in a distribution system, using annular reactors to represent water quality conditions in a distribution system and *E. coli* as a model organism. The experiments were conducted in two phases to attain these objectives. The experimental phases were *Phase 1* and *Phase 2. Phase 1* used laboratory studies to examine the effects of using single chemical disinfectants (i.e., free chlorine, chloramines and chlorine dioxide) alone or in combination with a UV treatment in both polycarbonate and cast iron pipes. The effect of the various disinefection schemes on *E. coli* mutations was also examined. *Phase 2* consisted of field studies of surface water (Halifax, Nova Scotia) and groundwater (Pinellas County, Florida) locations and examined the long-term biofilm ecology, and was evaluated for both genotypic and phenotypic changes.

CONCLUSIONS

UV treatment was extremely effective at eradicating *E. coli* prior to entering the model distribution system. In fact, synergistic effects for mitigating *E. coli* were observed when *E. coli* was spiked into the UV-reactor and then was treated with either ClO_2 or Cl_2. The benefits of combined disinfectants were only observed on *E. coli* K12, as the mut mutant was completely inactivated by UV-treatment. In fact, the mut mutant was not recovered from any AR following an initial UV treatment. It was also found that *E. coli K12* was much more resistant to UV disinfection than that the *mutS* mutant strain. The mutant strain was never cultured from the biofilm or effluent of any of the Post-UV ARs. In general, the laboratory experiments demonstrated that biofilm *E. coli* K12 was found to be be more resistant to disinfection than effluent *E. coli* K12. Very few isolates were obtained with the wild type *E. coli*.

The picture was somewhat different when the *E. coli* was added after the UV. Isolates were obtained in significant numbers when there was no disinfection (controls) and some isolates were obtained at low disinfectant concentrations typical of distribution system termini. Among these, there were antibiotic-resistant isolates for all treatments with a higher proportion of the isolates from chloramine treatment showing antibiotic resistant phenotypes to multiple antibiotics. The significance of antibiotic resistant phenotypes in drinking water is uncertain at this stage but clearly deserves further investigation both on the mechanistic side to determine the types of mutations involved and on the causative side to which is the most significant influence on the mutation rate - starvation and oxidative stress or the genotoxic effects of the disinfectant and or the DBPs.

Considering all observations from the two pilot studies and the laboratory study, it is concluded that the sequential inactivation of heterotrophic bacteria with UV light and secondary chlorine-based disinfectants shows synergistic benefits. In the controlled laboratory setting, synergistic effects were clearly observed for the inactivation of *E. coli.* In comparing disinfection strategies in field studies, combination treatments with UV as a primary disinfectant generally achieved highest log removals, and chlorine dioxide was the most effective disinfectant followed by chlorine and finally monochloramine. UV alone was not effective against heterotrophic bacteria and possibly increased bacterial regrowth potential in the ARs which led to increased HPC bacteria without secondary residual disinfection. Enhanced reduction with UV light in combination with secondary chlorine-based disinfection was observed in all water types regardless of quality or climate.

RECOMMENDATION

It is recommended that implementing UV light as a primary disinfectant with chlorine-based secondary disinfection will enhance removal of HPC bacteria due to synergistic effects and the previously established added benefit of a wider range of bacteria and protozoa inactivated (i.e. chlorine-resistant pathogens such as *Cryptosporidium parvum*). Utilities will be able to benefit from high reductions in HPC bacteria and lower formation of biofilm in the distribution systems. There would also be potential for lower CTs required for chlorine-based disinfectants with UV light pre-treatment to achieve higher reductions than chemical disinfectant alone at high CT. However, utilities would not be able to depend on UV pre-treatment to lower required dosages of chlorine-based disinfectant to maintain minimum residual concentrations. It is also recommended that UV light not be used as the primary and only disinfection because bacteria counts actually increased with UV-treated water when no residual protection was supplied.

To achieve downstream benefits of UV treatment, it is recommended that UV treatment proceeds as a secondary disinfectant. If UV follows chemical disinfection, the degradation of chemical residual in the presence of UV light produces negative results. Although studies have shown that pre-chlorination aids in controlling biological growth on the lamp itself, it is believed that this benefit is overshadowed by the negative impacts of pre-chemical disinfection (i.e. loss of Cl_2 residual). The reduction of residual due to UV treatment leads to increased bacteria counts because there is less available to control microbial regrowth in the distribution system. To combat this problem, utilities would need to additionally treat with chemical disinfection following UV irradiation in order to increase residual protection, which would potentially lead to higher disinfection by-product formation and increase chemical costs for disinfection. It is especially not recommended to implement secondary UV disinfection with monochloramine as a primary disinfectant, since it has been observed that nitrification occurred with this disinfection strategy leading to higher nitrate and nitrite concentrations.

The laboratory study presented in this project showed synergistic effects in reducing *Escherichia coli*, which is a pathogen easily reduced by UV light and chemical disinfection. It would be recommended to investigate if synergistic benefits still occur with a pathogen that would not be easily affected by UV light, and to determine if multiple damage mechanisms still enhance removal.

Furthermore, this project and other studies have shown that disinfection synergy occurs with sequential disinfection strategies. However, the mechanisms involved have not been identified. Molecular work in the present study indicates potential pathways for exploring the benefits of sequential treatment. However, further research is needed to more fully characterize possible mechanisms of synergy.

CHAPTER 1
PROJECT OVERVIEW

INTRODUCTION

This study aimed to significantly expand the current knowledge of the interaction and effect of UV and chemical disinfectants on *E. coli* in mixed species biofilms. The combined and individual effects of UV and chemical disinfectants were examined using using *E. coli* spiked into mixed species biofilm. Most studies on distribution system biofilms have used only culture to determine bacteria present. More recently applied molecular techniques (e.g. PCR-DGGE) offer some advantages in permitting determination of biofilm composition and dominant organisms, even if these cannot be isolated in culture (Muyzer, 1999).

Schwartz et al. (2003) used a combination of culture and molecular techniques to follow changes in biofilm composition with UV and ClO_2 application. Recording increased metabolic activity in biofilms treated with UV, and that ClO_2 was critical for mitigating pathogens such as enterococci although neither disinfection method could deal with more resistant pathogens such as legionellae and mycobacteria (Schwartz et al., 2003). No comparison of ClO_2, Cl_2 and chloramines was conducted in that study. Pryor et al (2003) used a combination of culture and molecular techniques to follow changes in biofilm composition with changes in disinfectant residual (chlorine and chloramine) and water source. The results showed some important changes in the opportunistic pathogens legionellae and mycobacteria. No UV disinfection was included in this study. Molecular techniques have also been used to follow composition and succession in biofilms in a model drinking water biofilm from 1 day to 3 years (Martiny et al., 2003). In addition a recent AwwaRF project, titled *Impact of UV Disinfection on Biological Stability* conducted by Gagnon et al. (2004), found that UV disinfection in combination with ClO_2 or Cl_2 resulted in an observed synergy, as quantified exclusively by heterotrophic bacteria using culture based methods. Synergistic effects between UV/ClO_2 and UV/Cl_2 resulted in greater disinfection in the distribution system than a single disinfectant (i.e., UV, ClO_2, Cl_2). It is hypothesized that molecular techniques will help to explain the synergies observed previously, particularly as they relate to potential pathogens.

In addition, emergence of antimicrobial resistance in bacteria, and its occurrence and or dissemination, results from the versatility of bacterial populations in adapting to less favourable and or toxic environments through changing and exchanging DNA. Broadly, antimicrobial resistance covers increased resistance to both disinfectants and antibiotics, although these may occur through similar or disparate cellular mechanisms. Bacteria in drinking water distribution systems encounter, or have encountered, a variety of stressful influences – reduced nutrient levels leading towards starvation, exposure to metals, particularly iron and manganese from pipe corrosion, the presence of sublethal levels of disinfectant residual(s), disinfection by-products and other potentially toxic chemicals in the water, and additional barriers such as UV that may be applied to maintain water quality. Adaptations to such stressors in drinking water are poorly understood but can generally occur at three levels that are not mutually exclusive:

- population – where the stress selects for those bacteria that are inherently more resistant
- physiologic – where the organism adapts through differential gene expression, and
- genetic – where random or adaptive mutation, or horizontal gene transfer, results in strains that are better fitted to the stressful environment.

Although it may often be difficult to distinguish the most significant level of adaptation, it is extremely valuable to know which pipe material or disinfectant residual is preferable in selecting or not for antimicrobial resistance in distribution systems. Thus, the proposed study would be the first to systematically examine the potential relationship between exposure of bacteria to disinfectants in drinking water and increases in antimicrobial resistance. Further, it will take a novel approach in looking also at the appearance of mismatch repair (MMR) mutants, because of their further potential for rapid change in water, the environment or human hosts. The MMR system most likely accounts for the increased levels of muations associated with the stationary phase indicating an adaptive mutation. MMR mutants occur naturally as about 1% of *E. coli* isolates. The one used here (mutS) is used simply as a means of comparing to WT to assess the degree of mutation and potential mechanisms of at least some of the resistance that develops. There is no particular significance of these types of mutants for the drinking water industry more than anywhere else that they occur. However, in at least one clinical study, significant numbers of urogenital *E. coli* isolates were shown to be mismatch repair mutants – mainly *mutS*. This is presumably because their high mutability allows them to adapt readily to the host – and potentially other situations.

RESEARCH OBJECTIVES

The major objective of this 24-month project was to understand the potential mechanisms of anti-microbial resistance of *E. coli* and opportunistic pathogens under disinfection conditions that are relevant to drinking water distribution systems. The research was conducted under controlled laboratory conditions and at the field level to provide closer approximations to the microbiological reactions in distribution systems. The project involved molecular techniques to understand genetic changes in biofilm and suspended bacteria in response to different disinfectant conditions.

Laboratory and field studies were performed in parallel to model pathogens in a distribution system, using annular reactors to represent water quality conditions in a distribution system and *E. coli* as a model organism. The experiments were conducted in two phases to attain these objectives. The experimental phases were as follows:

- *Phase 1:* Laboratory studies will examine a comprehensive experimental matrix, which will examine the effects of using single chemical disinfectants (i.e., free chlorine, chloramines and chlorine dioxide) compared with a UV pre-treatment, impact of pipe material (polycarbonate or cast iron) and *E. coli* mutations.
- *Phase 2:* Field studies at surface water (Halifax, Nova Scotia) and groundwater (Pinellas County, Florida) locations will examine long-term biofilm ecology and be evaluated using genotypic and phenotypic methods.

REPORT ORGANIZATION

The report includes general literature, and results from bench and field scale studies on the *E. coli* occurrences in drinking water distribution systems. In particular, the specific chapters address the following topics:

Chapter 2 provides an overview of biofilm formation and composition, coliform occurrence and associated health significance, various disinfection strategies, and the microbial stress response to stressors in drinking water.

Chapter 3 provides a detailed explanation of the materials and methods used throughout the experimental study. Design and operational parameters for both the bench-scale and field scale experiments are detailed, as well as relevant information about the system characteristics.

Chapter 4 evaluates the impact of the various disinfection schemes and pipe material on both heterotrophic plate bacteria and *E.coli* at the bench scale. The phenotypic response and antimicrobial resistance of *E.coli* is also evaluated at the bench scale. The field scale results are evaluated and a comparison of disinfectants on heterotrophic bacteria.

Chapter 5 evaluates the impact of disinfectant residual and upstream UV treatment on both effluent and biofilm *E. coli* in field scale study from a surface water source.

Chapter 6 evaluates the impact of disinfectant residual and upstream UV treatment on both effluent and biofilm *E. coli* in field scale study from a groundwater source in a warm climate.

Chapter 7 provides the overall project summary and recommendations to the water industry in terms of practical implications of the research.

CHAPTER 2
LITERATURE REVIEW

BIOFILM COMPOSITION

Reports by utilities show that microorganisms survive in distribution systems despite the continuing presence of disinfectants, thus placing doubt on the ability of a disinfectant residual to ensure protection from regrowth. In particular, a wide variety of bacteria have been isolated from distribution system biofilms. LeChevallier et al. (1990) reported isolating the following genera from distribution system biofilms: *Arthrobacter*, *Flavobacterium*, *Moraxalla*, *Acinetobacter*, *Bacillus*, *Pseudomonas*, *Alcaligenes*, and *Achromobacter*. *Arthrobacter* was also isolated from distribution system biofilms by Nagy and Olson (1985), who estimated that 20% of biofilm HPC bacteria were *Arthrobacter*. One of the main concerns associated with regrowth in distribution systems is the multiplication of potentially pathogenic bacteria. Camper et al. (1998) reported that opportunistic pathogens (*Klebsiella pneumoniae* and *Aeromonas hydrophila*) as well as frank pathogens (*Escherichia coli* O157:H7 and *Salmonella typhimurium*) were able to colonize surfaces and persist under conditions similar to drinking water distribution systems. Pathogens that have been isolated from distribution system biofilms include *Legionella spp.* (Schwartz et al., 1998a; Pryor et al., 2003), *Aeromonas hydrophila* (Chauret et al., 2001), *Pseudomonas spp.* (Ribas et al., 2000), *Mycobacterium* (Schwartz et al., 1998b; Falkinham et al., 2001; Pryor et al., 2003), and coliform bacteria (LeChevallier, 1990; Pryor et al., 2003).

Potential explanations for the occurrence of microorganisms in distribution systems include pathogen intrusion in the distribution system (Kirmeyer et al., 2001), chronic regrowth problems from poor treatment practices or by short-term process upsets. However, the persistence of microorganisms in distribution systems remains a long-term problem with limited explanation in the drinking water literature. It must be noted though that Ct value determinations are performed on suspensions of microorganisms, whereas there is ample evidence from microbial literature that the biofilm growth habit produces differential gene expression in biofilm organisms and that biofilm bacteria are significantly and intrinsically more resistant to antimicrobials than their planktonic counterparts.

COLIFORM OCCURRENCE/REGROWTH IN DISTRIBUTIONS SYSTEMS

Ensuring the microbial integrity of drinking water distribution systems is the final and one of the most critical steps in the multi-barrier approach for supplying safe drinking water. One of the most important monitoring parameters used in North America to ensure public safety against waterborne disease is the monitoring of total coliforms. The *Total Coliform Rule* issued by the United States Environmental Protection Agency (USEPA) and the *Canadian Drinking Water Guidelines* issued by Health Canada are in place to protect the public from outbreaks of waterborne disease.

Coliform occurrences in distribution systems occur as a result of inadequate treatment/maintenance of a disinfectant residual or because of an intrusion downstream from the plant (Craun & Calderon, 2001). In addition, coliforms may pass through treatment or persist in the distribution system undetected because they have been injured and are not detectable by currently accepted detection methods (McFeters *et al*, 1986a).

Coliform bacteria in drinking water can experience sublethal, reversible injury as a result of several chemical and physical factors. This form of injury is responsible for inhibiting their growth on accepted media used in microbial drinking water analysis (McFeters *et al*, 1995). Injured coliforms are susceptible to selective substances such as deoxycholate and bile salts that are commonly used in enumeration media such as m-Endo (LeChevallier *et a*l, 1983). The most prevalent factors that can induce stress on coliform bacteria in drinking water include: disinfectants such as chlorine, low concentrations of metals such as copper and zinc, limited nutrient supply, extremes in temperatures and pH and interactions with other microorganisms (McFeters *et al*, 1986a; McFeters *et al,* 1986b).

In a study conducted by LeChevallier and McFeters (1985), it was found that heterotrophic plate count (HPC) bacteria could reduce coliform densities. Organisms such as *P. cepacia* caused lethal interactions with coliforms while others such as *P. maltophilia* caused sublethal injury to coliforms (LeChevallier and McFeters, 1985). In addition, it was concluded that HPCs can grow at low nutrient levels, while coliforms have higher growth requirements. Consequently, coliforms are at risk because they must compete with other organisms for energy sources in low-nutrient environments such as drinking water (LeChevallier and McFeters, 1985). Starved coliforms are presumably more susceptible to other stress-inducing factors that can cause injury or death.

Initial results in a study performed by McFeters *et al* (1986) indicated that coliforms and enterotoxigenic *Escherichia coli* (ETEC) were much more susceptible than other organisms to both chlorine and copper injury. For instance, *Salmonella typhimurium* and *Shigella spp* required > 90% higher levels of chlorine and copper than the coliforms in order for injury to occur. As a result these "indicator bacteria" are much more susceptible to environmental factors in drinking water than common waterborne pathogens.

HEALTH SIGNIFICANCE OF COLIFORMS IN DISTRIBUTIONS SYSTEMS

Microbiological quality of water is determined by monitoring coliform bacteria, most notably those that ferment lactose and belong to the Enterobacteriaceae family (Singh & McFeters, 1992). Fecal coliform bacteria such as *E. coli* are not usually long term inhabitants of aquatic ecosystems as they originate primarily in mammalian intestines (Singh & McFeters, 1992). Consequently, these organisms serve as "indicator bacteria" of recent fecal contamination and may suggest that potentially harmful pathogens be present in the water.

According to the World Health Organisation (WHO), about 20% of the world's population does not have access to safe drinking water and more than 5 million people die annually from diseases associated with improper sanitation and/or unsafe drinking water (Hunter *et al*, 2001). According to Hunter et al. (2001), a variety of microorganisms can cause waterborne disease such as *Cryptosporidium*, *Cyclospora*, *E. coli* O157, *Legionella*, *Helicobacter pylori*, hepatitis E virus, *Toxoplasma*, and others.

Outbreaks pose a great risk to the public as certain groups of people such as children, elderly and immunoccompromised individuals are more sensitive to waterborne pathogens than others (Singh and McFeters, 1992). These people could die if infected by a pathogen while others belonging to the general population could simply experience symptoms of diarrhea. According to Singh and McFeters (1992), the elderly account for 50% of the deaths related to gastroenteritis and hepatitis A in developed countries. In the United States, through 1990-1998,

127 drinking water outbreaks were reported to the Centre for Disease Control (CDC) (Hunter *et al*, 2001). Examples of recent *E. coli* O157:H7 outbreaks in North America include:

- Walkerton, Canada in May 2002
    - 2300 people experienced gastroenteritis, 45 were hospitalized, 27 developed haemolytic uremic syndrome (HUS) and 7 people died (Hrudey et al, 2002)
- New York, USA in September 1999 (CDC, 1999)
    - 15,921 case of diarrhea, some stool samples showed co-infection with *Campylobacter jejuni*
- Cabool, USA in December 1989
    - 243 cases of diarrhea, 32 people were hospitalized and 4 people died (Geldreich et al, 1992)

These outbreaks were caused either by main breaks in distribution systems or by groundwater contamination. Fortunately over the past 20 years outbreak occurrences in the United States have been declining. This is attributed to the stringent regulations and monitoring requirements put forth by the USEPA along with the continuous efforts of water utilities and public health officials to protect public safety (Hunter *et al*, 2001).

DISINFECTION STRATEGIES

Final disinfection is the process of applying a disinfectant that provides a residual concentration in the distribution system and the primary mitigation step used by utilities for controlling microbiological growth in distribution systems.

The United States Environmental Protection Agency is presently reviewing regulations pertaining to final disinfection. Presently, the EPA requires that a maintenance of 0.2 mg/L disinfectant residual entering the distribution system (Federal Register document # 52 FR 42199). A minimum disinfectant residual of 0.2 mg/L is specified to assure confidence in the measurement. While it is recognized that 0.2 mg/L in the distribution system will limit growth of heterotrophic bacteria within the distribution system, the concentration of 0.2 mg/L provides a quantifiable minimum target. However there is relatively limited data that can specify the appropriate minimum disinfectant level for known pathogens in distribution systems, such as *E. coli*. Thus it is plausible that 0.2 mg/L may be the minimum detectable concentration, but the minimum concentration for inactivation could be different.

Ultraviolet Light

Microorganisms persisting in distribution system biofilms are already adapted to be more instrinsically resistant to disinfectant residuals. Giese (2002) examined the impact of UV disinfection, in the absence of chemical disinfection, on biofilm accumulation in a model distribution system and found that the impact of UV disinfection on DNA fingerprints of biofilm organisms was variable. This may be because of random changes (mutations) effected in microorganisms during exposure that may permit some organisms to flourish and others to perish.

Chemical Disinfectants

Free chlorine is the dominant choice for secondary disinfection for North American utilities (AWWA, 2000). However, it is well established that free chlorine can react with natural organic matter (NOM) to form harmful DBPs (i.e., trihalomethanes). Monochloramine can be a better biocide for controlling biofilm growth (Norton and LeChevallier, 1997). However, a concern with chloramination is nitrification, the microbiological conversion of ammonia to nitrite and nitrate. Wilczak et al. (1996) estimated that two thirds of medium to large U.S. utilities that chloraminate experience some nitrification.

Chlorine dioxide (ClO_2) is a strong disinfectant and oxidant that has demonstrated promise as a secondary disinfectant in full-scale distribution systems (Volk et al., 2002). The formation of organohalogens (e.g., trihalomethanes and haloacetic acids) with ClO_2 is typically much lower in concentration than when using free chlorine (Hofmann et al., 1999; Werdehoff and Singer, 1987). This is primarily due to the difference in oxidation reaction mechanisms between the two compounds, where ClO_2 is via free radical electrophilic abstraction and Cl_2 uses oxidative substitution and addition (Baribeau et al., 2002). Recent data presented by Gagnon et al. (2004) has demonstrated potential synergistic effects between UV and ClO_2 for controlling growth of heterotrophic bacteria in distribution systems. The synergistic was also observed for UV and Cl_2, but chloramines did not exhibit any potential advantages with UV.

MICROBIAL STRESSORS AND STRESS RESPONSES IN DRINKING WATER

More than 20 years ago Armstrong et al. (1981; 1982) documented increased levels of antibiotic resistant bacteria as a result of water treatment processes including disinfection. However, no further investigations followed to understand whether this was simply due to a selection of the most inherently resistant bacterial strains, or was a development of resistance in sensitive strains as a result of exposure to sub-lethal selection pressures, particularly disinfectants.

Disinfectant residuals exert strong selective pressures on bacterial flora. Under controlled conditions, sublethal exposures of bacteria to disinfectants can result in altered and often reduced susceptibility to other antimicrobials, including antibiotics (Gilbert and McBain, 2003). In addition to disinfectants and their byproducts, bacteria in drinking water are subject to several, almost continuous, ecological stressors that include: nutrient depletion/starvation, metals from pipe corrosion and microbial competition / predation. As an example, starvation has been clearly shown to affect the resistance of *E. coli* O157:H7 to chlorine (Maenhurt-Michel and Caillet-Fauquet 1982). Other genes are down regulated during starvation, and the global regulatory circuit controlled by *rpoS* is implicated (Hengge-Aronis, 2002). Resulting phenotypic changes, such as in broad spectrum efflux pumps or membrane porin protein alterations (Hancock and Brinkman, 2002) may account in large part for the increased bacterial resistance to antibiotics seen in treated water over raw water (Armstrong et al., 1981). Similar changes in gene expression are apparent in moving from planktonic to biofilm growth in *E. coli* (Schembri et al., 2003), and biofilm growth habit in *E. coli* also appears closely tied to nutritional status and central carbon flux control by CsrA (Jackson et al., 2002).

In addition, in distribution system conditions UV exposure is also known to cause mutations and, under specific conditions, it has been demonstrated that *E. coli* has the ability to repair and remain culturable following exposure to UV light (Zimmer and Slawson, 2002).

Collectively the microbial stressors in drinking water may affect the mutation rate of bacteria exposed to them and they could potentially act synergistically rather than in an additive fashion. Mutations that occur in the MMR genes, or at other genetic loci involved in monitoring replication fidelity, could result in 'mutator strains' (Kotewicz et al., 2003), that accumulate mutations because of non-functional repair mechanisms.

Random mutations, particularly of regulatory genes, such as *marR* (Martin et al., 1996), could result in acquisition of antimicrobial resistance by recognized mechanisms such as multidrug efflux pumps. Such broad spectrum efflux pumps have been the subject of several recent reviews and are aptly referred to as the bacterial "Vomit Response" (Gilbert & McBain, 2003) or "toxic waste disposal" (Helling et al., 2002). Recent characterization of the *mar* regulon (Barbarosa & Levy, 2000) shows that constitutive expression of *marA* altered (76% increased and 24% decreased) expression of more than 60 chromosomal genes in *E. coli* and confirm that the *mar* locus mediates a global stress response through one of the largest described gene networks. Recently, Potenski et al. (2003) implicated *mar* mutants resulting from exposure to chlorine concentrations relevant to the food industry.

There is also increasing recognition that mutational processes in bacteria are influenced by environmental stressors (Notley-McRobb et al., 2002). Of particular significance in the development and acquisition of antibiotic resistance may be the growth-related down regulation of the methyl-directed mismatch repair (MMR) system (Harris et al., 1997; Tsui et al., 1997). The MMR system most likely accounts for the increased levels of mutation associated with stationary phase – adaptive mutation. Mutations that occur in the MMR genes or at other genetic loci involved in monitoring replication fidelity could result in 'mutator strains' (Levy & Cebula, 2001), that accumulate mutations because of non-functional repair mechanisms. Different strains of *E. coli* can exhibit widely differing mutation rates and it is suggested that there is a high level of mutator strains among clinical isolates (Matic et al. 1997) but the data lack statistical significance. Under continuous selection pressure, the proportion of mutators can increase and even become dominant (Miller et al., 1999). Significantly, most naturally-occurring mutator strains isolated so far show defects in MMR genes (Kotewicz et al., 2003). It has recently been shown (Li et al., 2003) that the amount of *mutS* in wild-type *E. coli* O157:H7 decreased 26-fold in stationary phase compared with exponential growth; a much greater decrease than for *E. coli* K-12. *mutS* defects in natural isolates included extensive deletions and truncation. Of the four MMR genes, *mutS* mutants are by far the most common (Kotewicz et al., 2003). The strains used by Kotewicz et al. (2003) are the same ones we propose to use in this study.

It is believed (Taddei et al., 1997; Boe et al., 2000; Brown et al., 2001) that modulation of the MMR genes is an evolved means of promoting evolution and adaptation in bacteria and offers a means of avoiding the deleterious effect of mutations. Obviously, continuous mutation would soon lead to non-viability, but *mutS* mutants also show increased ability to accept DNA by recombination (Brown et al., 2001). Horizontal gene transfer by conjugation is known to occur in biofilms (Christensen et al., 1998) and potentially useful traits can be acquired from other biofilm bacteria. This is supported by clear evidence of gene acquisition in the large size variation in the *mutS-rpoS* intergenic region in enteric bacteria (Kotewicz et al., 2003). Moreover, Hausner and Wuertz (1999) have shown conjugation rates in biofilms to be 1000-fold greater than are measured by classical plating.

Mutator strains are relatively common among clinical isolates and comprise more than 1% of naturally-occurring environmental isolates (LeClerc et al., 1996). It is also probable (Tanabe *et al.*, 1999) that mutator strains infecting patients and then being exposed to antibiotics may rapidly acquire antibiotic resistance *in vivo*. Therefore a key question to be answered is whether exposures to sublethal levels of water disinfectants give rise to changes in the levels of *mut* mutants.

CHAPTER 3
MATERIALS AND METHODS

INTRODUCTION

This chapter will outline the experimental design, the model distribution system operation and characteristics as well as analytical methods used for both the bench scale and field scale experiments.

Laboratory and field studies were performed in parallel to model pathogens in a distribution system, using annular reactors to represent water quality conditions in a distribution system and *E. coli* as a model organism. Bench scale experiments were conducted at Dalhousie University laboratories, and DNA analysis was performed at the University of Ottawa. Most water quality analysis was performed using methods adapted from the *Standard Methods of Water and Wastewater Examination 21st Edition* (APHA 2005).

In Phase 1, laboratory studies examined a comprehensive experimental matrix, which will examine the effects of using single chemical disinfectants (i.e., free chlorine, chloramines and chlorine dioxide) compared with a UV pre-treatment, impact of pipe material (polycarbonate or cast iron) and *E. coli* mutations.

In Phase 2, field studies at surface water (Halifax, Nova Scotia) and groundwater (Pinellas County, Florida) location examined long-term biofilm ecology, and were quantified by both genotypic and phenotypic changes. The J.D Kline Water Supply Plant is used to treat raw surface water from Pockwock Lake in Halifax, Nova Scotia. The treatment process consists of direct filtration with chlorination for final disinfection. The source water is low in alkalinity (<10 mg/L as $CaCO_3$) and typically has no occurrences of *Cryptosporidium parvum* or *Giardia lamblia* since the Halifax Regional Water Commission owns the land surrounding their watershed. The Keller Water Treatment Plant in Pinellas County, Florida, uses ground water as well as a ground water and surface water blend with chloramines for final disinfection. From the 2001 Consumer Confidence Report, Pinellas County reported 0.3% of their samples positive for total coliforms and 0% positive for fecal coliforms.

PHASE 1 – BENCH SCALE EXPERIMENTS

Experimental Design

Annular reactors (ARs) (BioSurface Technologies Corps) were used to simulate conditions relevant to drinking water distribution systems. Twelve annular reactors were setup in parallel to address various factors that could influence biofilm growth in distribution systems. Six ARs contained polycarbonate coupons, while the other six contained cast iron coupons. Eight of the ARs were pre-treated with low pressure UV lamps (Trojan UV-Max), while four received no UV disinfection. All twelve ARs were chemically disinfected (secondary disinfection) with chlorine, chlorine dioxide or monochloramine depending on the experimental trial. *E. coli* was selected as a model organism to study biofilm ecology. Two strains of *E. coli* were used in the experiment: *E. coli K12* and an MMR Mutant to see whether or not the wild K12 strain would adapt better or worse than the mutant strain. *E.coli* K12 (MG1655) and *E.coli* TX3496 (*MG1655 mutS::Ω [Km^r; EcoRV]*) , a *mutS* mutant strain that was derived from wild type MG1655 by P_1 transduction i.e. *MG1655xP₁virTX3451*, were obtained from Dr. Winkler M.E. (Department of Microbiology and Molecular Genetics, University of Texas Houston Medical School, 77030-1501, USA) (Tsui et al., 1997). The *E.coli* ATCC 25922 strain, recommended by NCCLS, was used as an external control in antibiotic susceptibility tests and during screening for resistant mutants. The internal control was *E. coli* K12 (MG1655) that had not been exposed in DWDS biofilms.

The experimental setup addresses the factors presented in Table 3.1. The target disinfectant residuals for each disinfectant trial are presented in Table 3.2. These residuals are based on the disinfectant concentration in the effluent of the ARs.

Table 3.1
Experimental factors

Factor	Experimental levels
Pipe material	Polycarbonate or cast iron
Secondary disinfection	Chlorine (Cl_2) , chlorine dioxide (ClO_2) or chloramines NH_2Cl)
UV treatment	Presence or absence
E. coli strain	*E. coli* K12 (MG1655, wild type) or derived MMR mutant

Table 3.2
Target Disinfectant Residuals

Disinfectant	Target low residual (mg/L)	Target high residual (mg/L)
Chlorine	0.2	1.0
Chlorine Dioxide	0.2	1.0
Monochloramine	1.0	2.0

Annular Reactor Operation

The bench scale distribution systems used in this study were Model 1120 LS Laboratory Model Regrowth Monitors and Annular Reactors (AR) obtained from Biosurface Technologies Corporation (Bozeman, Montana). The annular reactor is comprised of a stationary outer glass cylinder that encompasses an inner rotating drum. The inner drum is controlled by a variable DC motor and can rotate between speeds of 25 to 430 rpm. The rotational speed of the inner drum is what simulates the shear stress encountered on a pipe wall in a distribution system. Vertical and horizontal mixing is achieved through four draft tubes that are found in the inner drum. The inner drum contains 20 flush-mounted coupons that contain the pipe material (polycarbonate or cast iron) upon which biofilm can accumulate. These coupons can be easily removed aseptically through the slide removal port for biofilm sampling. Influent water is pumped in through the influent port and effluent bulk water samples are taken from the effluent port. The hydraulic retention time (HRT) is controlled independently by the flowrate into the AR. In the reactor, water flows between the two cylinders. The volume of this annular gap is approximately 950mL.

For this study, twelve ARs were operated in parallel. They were operated at a rotational speed of 50rpm which translates into a shear stress of 0.25N/m^2 (Camper 1995). A shear stress of 0.25 N/m^2 corresponds to a flow of approximately 0.30m/s (1fps) in a 100mm (4in) diameter smooth pipe which is similar to shear conditions of other pilot and bench scale investigations of drinking water distribution systems (e.g. Camper 1996; Gagnon et al 2004; Sharp et al 2001).

The HRT for the reactors was six hours which translated into a flowrate of 2.83mL/min into the ARs. An HRT of 6 hours was chosen for several reasons:

1) Simulate later sections of a distribution system
2) Allow enough time for the *E. coli* to establish itself within the biofilm of the reactors

All glass surfaces and any exposed surfaces were covered with aluminum foil to reduce the potential for phototrophic growth within the system. All tubing feeding influent water, *E. coli* and disinfectant into the system were non-transparent. The experiments were conducted at a water temperature of 20 +/- 1°C.

Before each experimental trial, all annular reactors were completely disassembled and cleaned with antibacterial soap. All the coupons were removed from the drums and individually cleaned. Once reassembled, the ARs were filled with 70% ethanol solution and allowed to soak for 24 hours. All the tubing from the previous trial was discarded except the tubing used to feed disinfectant. New tubing that had been flushed with 70% ethanol, followed by de-ionized water was used for feeding influent water throughout the experiment and for the *E. coli* spikes.

Bench Scale Distribution System Setup

A general schematic of the bench-scale distribution system setup is presented in Figure 3.1. Tap water from the Halifax Regional Water Commission (HRWC) was the primary source water for the model distribution systems. Prior to entering the annular reactors, tap water passed through a granular activated carbon (GAC) filter to neutralize any chlorine present in the source water. Then the water was passed through a biologically active (BAC) filter which provided an inoculum for the ARs. The water from the BAC filter was then directed into a clearwell. The flow from this clearwell was split to feed both the UV unit and four annular reactors. Once water passed through the UV system it was directed towards another clearwell which fed eight ARs.

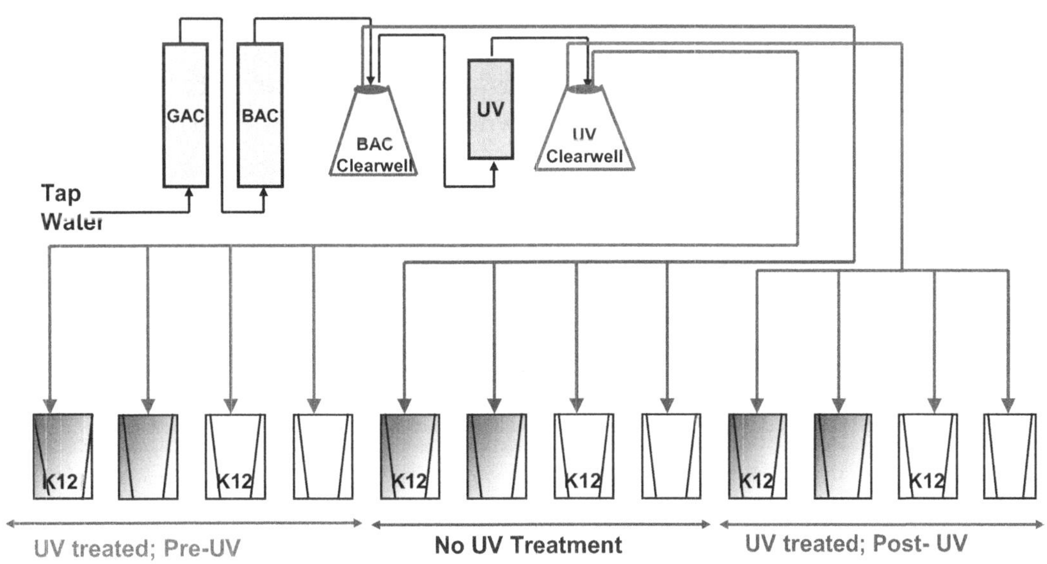

Figure 3.1 Bench Scale setup

The bench-scale setup can be broken up into three groups of four ARs. The groups were divided up by treatment method: Pre-UV, Non-UV and Post-UV (Figure 3.1). Each of these groups contained four ARs operating in parallel: two containing cast iron coupons and two containing polycarbonate coupons (Figure 3.1). In addition, two ARs of each group were spiked with *E. coli* K12 and the other two were spiked with a MMR mutant strain of *E. coli* K12. Figures 3.2-3.4 demonstrate the difference between treatment methods. Figure 3.2 outlines the treatment regime for the Pre-UV treated ARs. The term "Pre-UV" means that UV disinfection occurred prior to the *E. coli* spike, therefore one of the two strains of *E. coli* was spiked directly into the annular reactor. These "Pre-UV" ARs were spiked after UV disinfection to simulate a breech in distribution system integrity.

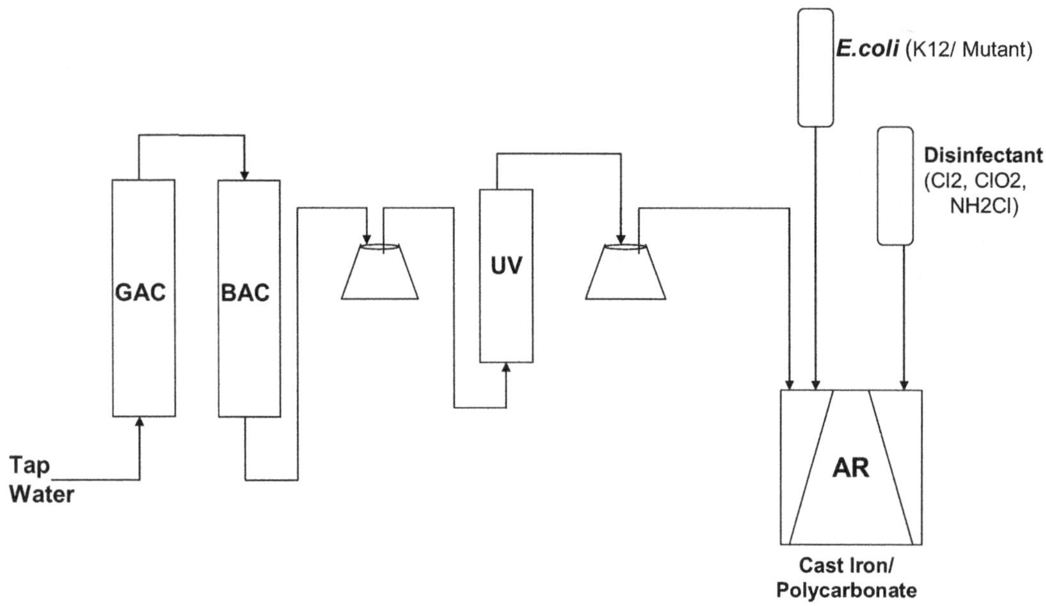

Figure 3.2 Pre-UV treatment setup

Figure 3.3 illustrates the setup for the second group of ARs that receive No-UV treatment. These ARs were also directly spiked with *E. coli* like the "Pre-UV" treated ARs.

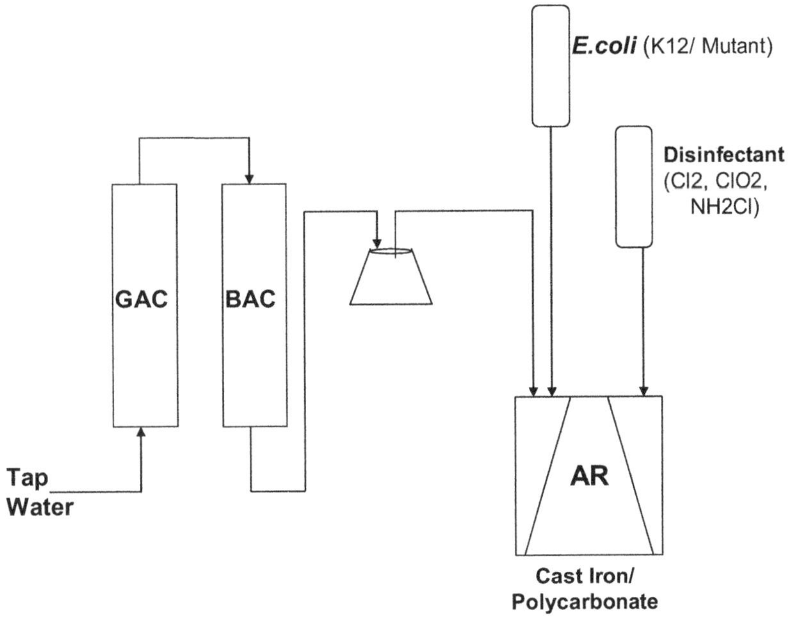

Figure 3.3 Non-UV treatment setup

Finally, the treatment schematic for the third group of annular reactors is presented in Figure 3.4. "Post-UV" treatment means that UV disinfection occurs following the *E. coli* spike. These "Post-UV" treated ARs were spiked with *E. coli* prior to disinfection to evaluate synergies that may occur through sequential disinfection. All three groups of annular reactors were disinfected with the same secondary disinfectant (Cl_2, ClO_2, and NH_2Cl) depending on the experimental trial.

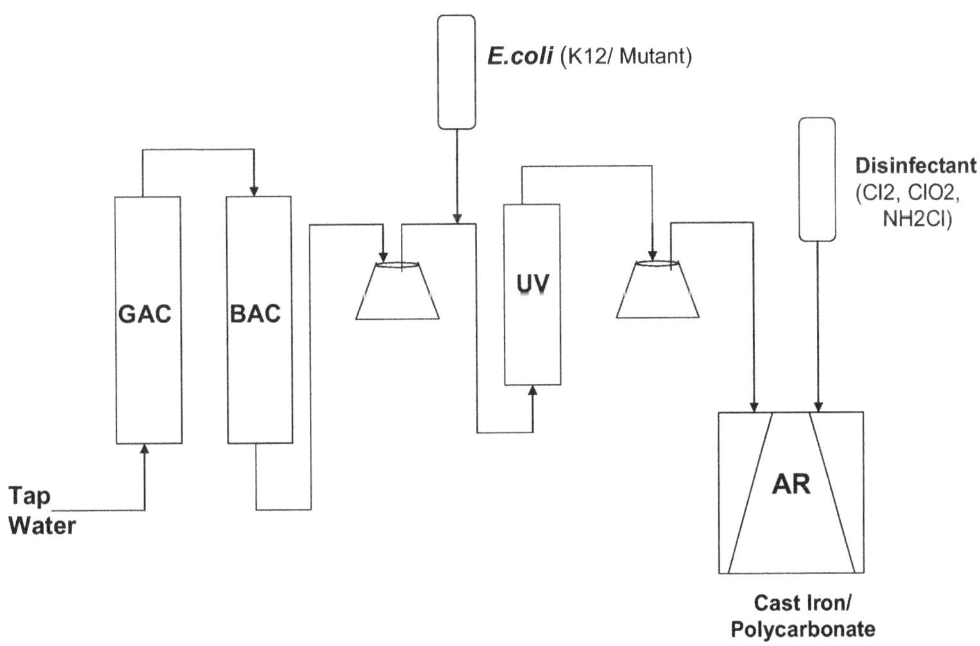

Figure 3.4 Post-UV treatment setup

Experimental Timeline

Four experimental trials (runs) were conducted over a period of 14 months. Each trial was conducted with a separate secondary disinfectant. Runs 1 & 2 were conducted with free chlorine at the secondary disinfectant. Run 3 used chlorine dioxide while run 4 looked at chloramines as a secondary disinfectant. Each trial followed the same experimental timeline. ARs were given three to four weeks to form an indigenous biofilm and reach steady state. At this point they were inoculated with one of two strains of *E. coli* K12 (ATTC strain #29425 and MMR Mutant). After a four week inoculation period, low level disinfection (0.2mg/L residual Cl_2, ClO_2; 1.0 mg/L residual NH_2Cl) proceeded for four to five weeks followed by high level disinfection (1.0mg/L residual Cl_2, ClO_2; 2.0 mg/L residual NH_2Cl) for three weeks.

Bench –Scale Distribution System Characteristics

Raw Water

Tap water from the Halifax Regional Water Commission (HRWC) was the source water used to feed the model distribution systems. Surface water from Pockwock Lake undergoes conventional treatment at the J.D. Kline water treatment plant. On average the water quality of this source water leaving the plant has an alkalinity of 17.5 mg/L as $CaCO_3$, a pH of 7.2 and a total organic carbon concentration (TOC) of 1.5mg/L (HRWC, 2005).

GAC/ BAC Filters

Tap water was fed up flow through a granular active carbon (GAC) filter to remove any free chlorine present in the water. Following this GAC filter, the water was fed up flow through a biologically active GAC column (BAC) to remove any natural organic matter (NOM) from the tap water.

UV System

A low pressure UV lamp (TrojanUV Max Model C) provided by Trojan Technologies was used throughout the experiments. This system is capable of treating up to 64 lpm (17gpm) at a fluence of 16mJ/ cm^2 and a UV transmittance (UVT) of 95%. Water was pumped through the lamp with a variable speed modular peristaltic pump (Masterflex) at a flowrate of 250mL/min which translated to a fluence of between 90 -100mJ/cm^2. Prior to the startup of each trial, the UV lamp was cleaned and soaked in 70% ethanol for 24 hours. The fluence was determined by performing chemical actinometry using potassium ferrioxalate (Dykstra et al. 2002).

Influent Water Characteristics

Prior to entering the bench- scale systems the water quality characteristics after passing through the GAC/BAC system and UV system are summarized in Tables 3.3 and 3.4, respectively.

Table 3.3
Average Water Quality for the GAC/BAC treated water

Water Quality Parameter	Average Measurement
pH	7.07
Turbidity	0.31 NTU
Temperature	19.4 °C
Total Organic Carbon (TOC)	1.930 mg/L
Dissolved Organic Carbon (DOC)	1.639 mg/L
UV 254	0.0262
Free Chlorine	n.d.*
Heterotrophic Plate Counts (HPC)	3.57 x 10^4 CFU/ mL
Total Coliforms (TC)	0 CFU/ 100mL

**n.d. – non detect (below detectable range)*

Table 3.4
Average Water Quality for the UV treated water

Water Quality Parameter	Average Measurement
pH	7.04
Turbidity	0.33 NTU
Temperature	19.5 °C
Total Organic Carbon (TOC)	2.064 mg/L
Dissolved Organic Carbon (DOC)	1.655 mg/L
UV 254	0.0212
Free Chlorine	n.d.*
Heterotrophic Plate Counts (HPC)	1.05 x 10^4 CFU/ mL
Total Coliforms (TC)	0 CFU/ 100mL

**n.d.- non detect*

E. coli Spike Procedure

E. coli K12 (ATTC strain #MG 1655) and the Mutant were obtained from the Centre for research in Environmental Microbiology at the University of Ottawa. Pure cultures were aseptically streaked onto Tryptic Soy Agar (TSA) and incubated for 2-5 days at 37.5 °C. For each spike, 24 plates were streaked with pure culture of either the wild or mutant *E. coli* K12 strain. Any extra plates were sealed in plastic bags and refrigerated at 4 °C for up to a month. All colonies from the plates were removed aseptically using a sterile loop and placed into sterile tubes containing 10mL of Tryptic Soy Broth (TSB). These tubes were then re-incubated for 24 hours at 37.5 °C. Following incubation, the tubes were centrifuged at 4000rpm for 10 minutes to separate the supernatant from the *E. coli* pellet. The supernatant was discarded and 9 mL of sterilized sodium chloride solution was added to the tube (0.85% w/v). The tubes were vortexed to re-suspend the pellet and "wash" the organisms in the salt solution. At this time the tubes were centrifuged again at 4000rpm for 10minutes. This washing process was repeated twice. Four tubes of the washed *E. coli* mixture were added at a time to 4L amber bottles containing sterile 0.85% sodium chloride solution (w/v). Four tubes of *E. coli* K12 per 4L amber bottle achieved a concentration of approximately 10^7- 10^8 coliforms/ mL. Six bottles were made for each spike, therefore one bottle for every two annular reactors.

The *E. coli* spikes during each trial took place over a couple of days. First, the Non-UV and Pre-UV treated ARs were spiked continuously for 24 hours. The amber spike bottles were setup on magnetic stir plates to ensure that the coliforms were continuously mixed during the spiking period. Influent flow into the ARs was shut-off during the spike to prevent flushing the *E. coli* out of the system. After the 24 hour period, effluent water samples were analyzed for total coliforms and *E. coli* to verify that the spike was successful. At this point, the flow was restarted to the ARs.

Following the first spike, *E. coli* was spiked through the UV lamp for the Post-UV treated ARs. During this spike, the influent flow into the Pre-UV treated ARs was switched from UV water to GAC water in order to avoid cross contamination of the systems. The spike bottles were fed simultaneously through the UV lamp with GAC treated water at approximately 10% of the total flowrate. The influent concentration of *E. coli* into the UV lamp was between 10^5- 10^6 CFU/mL. The mutant and wild *E. coli* strains were spiked separately through the UV lamp. In between all spikes through the UV lamp, the lamp was disinfected with 70% ethanol and during that time the influent water into the ARs was switched to GAC water.

Disinfectant Preparation

Chlorine

Chlorine stock bottles were prepared in 4 litre amber glass bottles containing de-ionized water. The bottles were autoclaved at 121°C for 30 minutes. Appropriate amounts of concentrated sodium hypochlorite (60,000mg/L OCl^-) were added to produce stock bottles at concentrations of 30-40mg/L (for polycarbonate ARs) and 120-140mg/L (for cast iron ARs).

Chlorine Dioxide

Chlorine dioxide was generated using a bench-scale procedure adapted from the *Standard Methods for the Examination of Water and Wastewater, 19th ed.,* a method used by Sterling Pulp Chemicals, Ltd. (Toronto) and one used by the University of Toronto- Pulp and Paper Research Institute. Chlorine dioxide generation took place in a fume hood using a bench scale apparatus. Chlorine dioxide gas was generated by dripping 25% sodium chlorite solution ($NaClO_2$) into an 18 N solution of sulfuric acid (H_2SO_4). The gas was purged off this reaction (Bottle 1) into a second gas washing bottle containing more $NaClO_2$ (Bottle 2). The chlorine gas contamination purged off Bottle 1 reacts with the $NaClO_2$ in Bottle 2 to produce more chlorine dioxide. The essentially "pure" chlorine dioxide is then purged off Bottle 2 into Bottle 3 where it is then trapped in deionized water. Bottle 3 is surrounded in ice to prevent volatilization of the ClO_2. Any chlorine dioxide gas that is not trapped in Bottle 3 is directed to a Potassium Iodide (KI) trap (Bottle 4) which neutralizes the gas before being let out into the air. Gases are purged from one bottle to the other using a vacuum source.

Monochloramine

Monochloramine stock was generated as follows: first 100mL of Phosphate Buffer Solution (PBS) was diluted into 900mL of distilled water. The pH was adjusted to 9.4 using sodium hydroxide. Ammonium chloride was produced by adding 4 g of ammonium chloride to

500 mL of the diluted PBS solution. Chlorine stock was produced by adding 18.7 mL of sodium hypochlorite (60 000 mg/L OCl^-) to the other 500 mL of diluted PBS. The chlorine stock was then mixed into the ammonium chloride solution using a magnetic stirrer and stir plate. This method produced a 1 L monochloramine stock solution of between 2800 mg/L and 3500 mg/L. Monochloramine stock solution was added to sterile 4 L amber bottles containing de-ionized water, and pumped to ARs to achieve the necessary residual concentration.

PHASE 2 - FIELD SCALE EXPERIMENTS

Experimental Design

Similarly to bench-scale studies, annular reactors (ARs) were used to simulate drinking water distribution systems in both field experiments. The set-up in Halifax consisted of six ARs in parallel receiving either raw surface water (RW) or raw water pre-treated with Ultraviolet light (UV). Chemical disinfection schemes included chlorine and chlorine dioxide following UV pre-treatment. A three-week acclimation period that allowed biological steady-state conditions to be reached within the ARs was followed by disinfection.

The set-up in Pinellas County, Florida differed from bench-scale and the Halifax experiments because the water sources for this plant are pre-treated with either chlorine or monochloramine. Therefore, there was no acclimation period and also UV treatment followed chemical disinfection. Disinfectant residual was the result of the concentration within the source waters as no additional chemical disinfectant was fed to the experiment system. The study ran over a seven-month period from April to November, 2005.

Annular Reactor Operation

The model used for both field operations was the same as used in bench-scale studies. In Halifax, each of the six ARs contained coupons made of polycarbonate. The hydraulic retention time (2.0h) was controlled by the volumetric flow rate of the influents entering the AR and was established based on a relationship between surface areas and volumes in ARs and pipes, as described by Gagnon and Huck (2001). Similarly to previous studies, the chemical flowrate was established at 5% of the total flowrate, which corresponded to 0.4mL/min for disinfectants and 7.7 mL/min for influent source water. The ARs operated at a rotational speed of 50rpm which translates into a shear stress of $0.25N/m^2$ (Camper 1995). A shear stress of $0.25N/m^2$ corresponds to a flow of approximately 0.30m/s (1fps) in a 100mm (4in) diameter smooth pipe which is similar to shear conditions of other pilot and bench scale investigations. One AR acted as a control for each water source (raw and UV-treated) and received no chemical disinfection. One RW AR and one UV AR were disinfected with chlorine, and one of each was disinfected with chlorine dioxide. The target disinfection residual concentration was 0.2 mg/L for each chemical, which is the minimum allowed in distribution systems under Nova Scotia drinking water regulations. A three-week acclimation period to allow biofilm to reach steady-state conditions in all ARs was followed by chemical disinfection for a total of 12 weeks.

The ARs set up in Florida contained coupons made of PVC material and operated under similar conditions at 50rpm with a HRT of 2.0h. Two ARs acted as controls and received raw source water containing free chlorine or chloramine residual. Two more ARs received water with either Cl_2 or NH_2Cl and were additionally treated with UV light. No chemical disinfectants were fed into the system to increase or decrease residual concentration, therefore no goal residual was targeted.

All non-opaque exposed surfaces of the ARs were covered to reduce the potential of phototrophic growth in the field systems. Before each experimental trial, all ARs were cleaned with antibacterial soap and disinfected using a 70% ethanol solution. In addition all tubing and clearwells used within the set-up were disinfected with ethanol for a period of 24 hours. This was followed by rinsing with Milli-Q water and the source water.

Bench Scale Distribution System Setup

A general schematic of the Halifax distribution system setup is presented in Figure 3.5. Surface water from Pockwock Lake was the primary source water for the model distribution systems. The water collected from a raw water tap was directed into a RW clearwell. The flow pumped from this clearwell was split to feed both the UV unit and three ARs. Once water passed through the UV system it was directed towards another clearwell which fed three additional ARs.

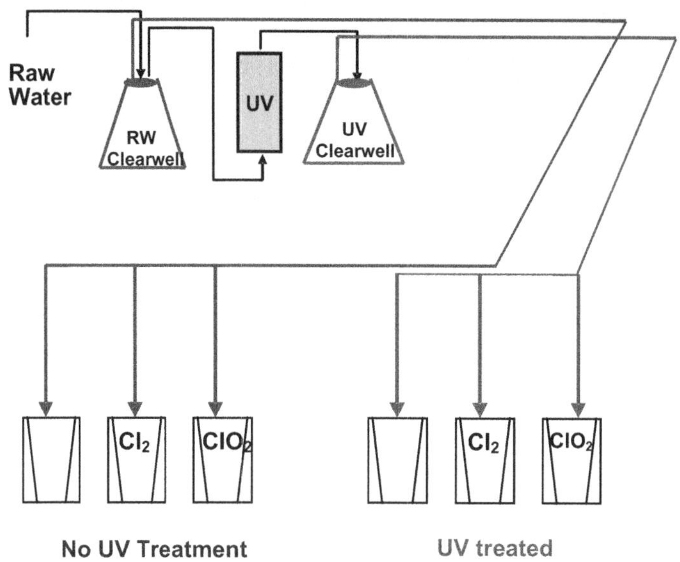

Figure 3.5 Schematic of the Halifax Field Scale Setup

The ARs were monitored 1-2 times per week for heterotrophic bacteria counts (suspended and biofilm), pH, temperature, UV254, turbidity, disinfectant residual, TOC, coliforms, and total trihalomethanes (TTHMs) using Standard Methods (APHA, 1995). In addition, weekly checks of flowrates throughout the system and the rotational speeds of the ARs were performed to ensure consistent operating conditions.

A general schematic of the Florida distribution system setup is presented in Figure 3.6. Two streams, one containing monochloramine and one with free chlorine, were the primary source waters for the model distribution systems. The water collected from each stream was directed into a separate RW clearwell. The flow pumped from each of these clearwells was split to feed both a UV unit and one AR. Each water stream had a separate UV system. Once water streams passed through the UV treatment they were directed towards two additional separate clearwells which fed the remaining two ARs.

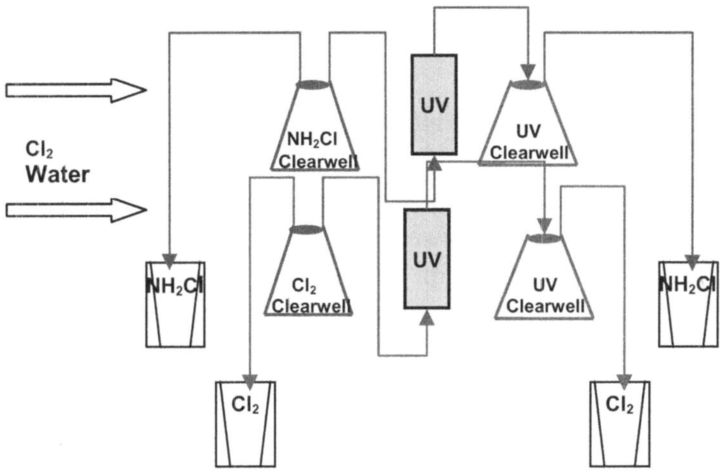

Figure 3.6 Schematic of Pinellas County Field Scale Setup

The ARs were monitored once weekly for heterotrophic bacteria counts (suspended and biofilm), pH, temperature, turbidity, disinfectant residual, TOC, coliforms, nitrites, nitrates, free ammonia, and total ammonia. In addition, weekly checks of flowrates throughout the system and the rotational speeds of the ARs were performed to ensure consistent operating conditions. Finally, three times over the course of the experiment samples were analyzed for total trihalomethanes.

Experimental Timeline

The experiment in Halifax consisted of two runs over a four month period from April to July 2005. Each run consisted of a 3-week acclimation period and was followed by six weeks of disinfection with goal residual of 0.2 mg/L. The Pinellas County study consisted of one run over a seven month period with a consistent disinfection strategy throughout the entirety of the experiment.

Field Scale Water Quality

Raw Water

The source water for the Halifax experiment was raw surface water from Pockwock Lake. This was directly applied to the ARs or pre-treated with UV light then directed into ARs. The water quality characteristics of each influent water stream are presented in Tables 3.5.

Table 3.5
Average Water Quality for the Raw and UV Treated Water

Water Quality Parameter	Average Measurement	
	Raw Water	UV Treated Water
pH	5.00	5.01
Turbidity	1.94 NTU	1.81 NTU
Temperature	19.4 ºC	19.9 ºC
Total Organic Carbon (TOC)	3.73 mg/L	3.51 mg/L
Dissolved Organic Carbon (DOC)	3.02 mg/L	3.27 mg/L
UV 254	0.109	0.104
Heterotrophic Plate Counts (HPC)	5.27 x 10^3 CFU/ mL	1.50 x 10^4 CFU/ mL
Total Coliforms (TC)	0 CFU/ 100mL	0 CFU/ 100mL

The two source waters for the Pinellas County setup included a chlorinated ground water and a ground water and surface water blend treated with chloramines. Each was applied directly to an AR or treated with UV then directed to additional ARs. Water quality characteristics for each of the four AR sources are presented in Table 3.6.

Table 3.6
Average Water Quality for the Raw Chloraminated and ChlorinatedWater

Water Quality Parameter	Average Measurement	
	Raw Chloraminated	Raw Chlorinated
pH	7.80	7.73
Turbidity	2.12 NTU	0.50 NTU
Temperature	26.1 ºC	26.2 ºC
Total Organic Carbon (TOC)	3.85 mg/L	3.70 mg/L
Monochloramine	1.67 mg/L	0.53 mg/L
Total Chlorine	2.13 mg/L	0.68 mg/L
Heterotrophic Plate Counts (HPC)	1.09 x 10^6 CFU/ mL	4.20 x 10^2 CFU/ mL
Total Coliforms (TC)	0.87 CFU/ 100mL	0 CFU/ 100mL

SAMPLE COLLECTION AND ANALYSIS

Microbial Enumeration and Characterization

Bench Scale

During the acclimation and spiking phases of each trial, weekly effluent and biofilm samples were taken from each AR. Weekly samples were also taken of the raw tap water, and influent water into the ARs (i.e. UV and GAC/BAC water). During both low and high disinfection phases sampling increased to 2-3 times/ week. Microbial analysis of these samples included heterotrophic plate counts (HPCs) and coliform/ *E. coli* analysis.

Effluent samples were collected in sterile, 15mL disposable plastic tubes (Corning Inc., Acton, MA). During the disinfection phases, residual disinfectant was quenched by adding 10µL of sterile sodium thiosulfate (10% w/v) solution to each tube.

Biofilm samples were collected aseptically by removing one coupon (per sample day) from each annular reactor. The coupons were pulled out of the reactors through the slide removal port using a sterile hook and were transferred immediately into 50mL glass tubes containing 25 mL of sterile PBS solution. During disinfection, 25μL of sterile sodium thiosulfate (10% w/v) solution was added to each glass tube. The biofilm was removed from the coupon surface by one of two methods: manual scraping or stomaching. Attached cells were removed from the cast iron coupons by manually scraping the coupon surface into a sterile stomacher bag (177 x 302 mm). The 50 mL of PBS solution was used to rinse the remaining biofilm off the coupon and into the stomacher bag. Prior to use, the hook and blade used to scrape and hold the coupon were sterilized with 70% ethanol and flamed three times. Biofilm was removed from the polycarbonate coupon surface using a MIX 2 Stomacher (AES Laboratories). The polycarbonate coupon was transferred from the 50mL test tube into a sterile stomacher along with the 25mL of PBS solution. The bag was then placed in the stomacher which operated at a speed of 230rpm for a period of 2 minutes (Gagnon and Slawson 1999). The liquid suspension (for both polycarbonate and cast iron coupons) was transferred aseptically from the stomacher bags back into the original sterilized 50mL glass tube. Following biofilm removal, all coupons were cleaned with antibacterial soap and sterilized with 70% ethanol and aseptically replaced back into the annular reactors. The inner drums of the ARs were labeled accordingly in order to prevent sampling the same coupon twice.

Field Scale

For the Halifax study, effluent, biofilm and influent samples were collected once weekly during acclimation and twice weekly during disinfection. Microbial analysis of these samples included heterotrophic plate counts (HPCs) and coliform/ *E. coli* analysis.

Effluent samples were collected in sterile, 50mL disposable plastic tubes (Corning Inc., Acton, MA). During the disinfection phases, residual disinfectant was quenched by adding 10μL of sterile sodium thiosulfate (10% w/v) solution to each tube. Biofilm samples were collected when the polycarbonate coupons were removed aseptically, in sequence from the AR. Once taken out, they were transferred into sterile 50-mL test tubes containing 25 mL autoclaved phosphate buffered saline (PBS) and 0.1% w/v sodium thiosulfate. The attached cells were immediately removed by the scraping method as described by Gagnon and Slawson (1999), and plated on agar to be analyzed for heterotrophic plate counts. Following sampling, the coupons were cleaned with soap and water, and treated with ethanol before being returned to the AR.

Microbial analysis for the Florida study was done in the Dalhousie University laboratory facilities and therefore samples required overnight shipment. Bulk samples were collected in 100mL IDEXX bottles containing 10% w/v sodium thiosulfate to quench disinfectant residual. The PVC coupons were removed aseptically and placed in sterile 50mL glass containing PBS and 0.1% w/v sodium thiosulfate. Once shipped, the coupons were scraped in the same method as the Halifax study to remove attached cells. New coupons that were treated with ethanol were used to replace those that were removed and shipped from each AR.

Heterotrophic Plate Counts

Heterotrophic plate counts were performed on R2A agar (Difco Laboratories) using standard spread plate techniques as described in *Standard Methods for the Examination of Water and Wastewater* (21st edition). Serial dilutions were prepared using glass test tubes containing 9mL of sterile phosphate buffer solution. Dilutions were used to target a microbial yield of between 30 to 300 colonies per plate. To ensure quality assurance, duplicate plates were spread for each dilution and all dilutions and plating occurred on a clean surface and close to an open flame. Plates were incubated upside down and in the dark for 7 days at room temperature and then all colonies were counted and recorded.

Coliforms/ E. coli

Bench Scale. Coliforms/ *E. coli* were enumerated from biofilm samples using the standard membrane-filtration technique for total coliforms as outlined in *Standard Methods for the Examination of Water and Wastewater* (21st edition). Appropriate serial dilutions were prepared using glass test tubes containing 9mL of sterile phosphate buffer solution. Samples were filtered aseptically through sterile 0.45μm filters using a vacuum aspirator. The filters were then transferred using sterile forceps to tight lid plates containing pre-dried m-Endo Agar LES (Difco Laboratories). These plates were then incubated upside down at 37.5 °C for 22- 24 hours after which all positive coliforms were enumerated (all colonies with a metallic- gold/green sheen). Since there were no coliforms in the influent water into the model distribution systems, all colonies counted on m-Endo Agar LES were considered to be *E. coli.* All filtration equipment and rinse water was autoclaved at 121°C for 15 minutes prior to use.

Coliforms/ *E. coli* were enumerated in AR effluent using the IDEXX Colilert® Quanti-tray® system. Sterilized bottles were used to collect 100mL effluent samples from the ARs. During disinfection these samples were quenched with 100μL sterile sodium thiosulfate (10% w/v) solution. Once the samples were collected, one Colilert® reagent packet was added to each sample and shaken until all the packet had dissolved. Then the samples were transferred from the 100mL sample bottles into the IDEXX Quanti-trays® and sealed used the Quanti-tray® Sealer. The trays were incubated for 24 hours and then counted using a Most Probable Number (MPN) table provided by IDEXX.

Field Scale. Coliforms/ *E. coli* were enumerated in AR effluent using the IDEXX Colilert® Quanti-tray® system. Sterilized bottles were used to collect 100mL effluent samples from the ARs. During disinfection these samples were quenched with 100μL sterile sodium thiosulfate (10% w/v) solution. Once the samples were collected, one Colilert® reagent packet was added to each sample and shaken until all the packet had dissolved. Then the samples were transferred from the 100mL sample bottles into the IDEXX Quanti-trays® and sealed used the Quanti-tray® Sealer. The trays were incubated for 24 hours and then counted using a Most Probable Number (MPN) table provided by IDEXX. Both field studies analyzed for coliforms onsite and no shipment of the Pinellas County samples was necessary.

API 20E

API 20E test kits were performed on *E. coli* colonies recovered from the biofilm during each experimental phase (*E. coli* Acclimation, Disinfection) in bench scale experiments. Duplicate samples (if available) of *E. coli* colonies from each annular reactor were tested using the API 20E test strips to see if the *E. coli* exhibiting any phenotypic changes from phase to phase.

Biofilm Community Composition

Determining the composition of biofilms in drinking water distribution systems, or elsewhere, is not an easy task, and necessarily can be relatively site specific as well as being subject to temporal and spatial variation. Nevertheless, there are likely to be a lot of commonalities. There are two basic approaches that can be used – either culturing the microbial cells directly and using a variety of identification techniques, or extracting the DNA and identifying bacteria following amplification of specific gene sequences for which much sequence information is available. Each of these methods provides some bias to the data obtained. When cells are cultured, only those microbial cells that are capable of using the substrates provided in the growth media will be recovered, though there may be 'hangers-on' that can benefit from metabolites produced by those microorganisms that grow. When DNA is extracted and amplified by polymerase chain reaction (PCR) there are likely to be unknown biases that favour extraction or amplification of certain species over others. Since the biases for the molecular methods are unknown, a common assumption is that the organisms detected are likely to be among the more numerically dominant. In some cases this is undoubtedly true but unlikely to be universally so. We chose to use employ both approaches simultaneously, to try and arrive at a broad spectrum of the bacteria present, though each approach had to be limited for practical reasons. Cultured bacteria were limited to those that would grow on R2A agar. Since these are not necessarily readily identified biochemically, and because we wanted to include even those on overgrown plates, we used sweeps of the R2A plates for further examination by molecular methods. The molecular method employed was PCR-DGGE using universal primers for the 16S ribosomal RNA gene. In this study we did not look for the presence of archaea or eukaryotes, though either or both of these may also be present.

Sample Types and Shipping

Samples for the biofilm analysis were obtained from the annular reactor coupons in Halifax. Samples received in Ottawa were approx. 10mL. Samples of the effluent from the annular reactors (approx 30 mL) were also collected. Heterotrophic bacteria in various dilutions were plated on R2A agar. All three types of samples were shipped in coolers with ice packs to Ottawa from Halifax by overnight courier.

Upon arrival, samples were inspected for leaks and/or contamination due to any mishandling during shipping. Difficulties noted included:

a) loss of effluent samples from AR1 and AR7 for the high Cl_2 treatment (1st run) due to the tube caps coming open during shipping;

b) this also contaminated one of the AR4 effluent R2A plates and one of the AR1 biofilm R2A plates, however since there were two other dilutions for each, those two plates were discarded without any major consequence;

c) AR12 for the pre-treatment Cl_2 (2nd run) was not ready, and therefore no samples (biofilm, effluent, or plates) were obtained;

d) UV effluent R2A plate for the high ClO_2 run was not shipped to the University of Ottawa team for processing;

e) all of the pre-treatment samples for the chloramine run were unfortunately left sitting at room temperature for one day during shipping since FedEx believed the shipment to be "dangerous goods" and returned them to sender.

Sample Preparation and Processing

Before proceeding to isolate the DNA, each sample was vortexed to mix, and a small quantity was retained as a back-up in a screw cap tube stored at 4 °C. When possible, biofilm and effluent samples were processed immediately, however due to the number of samples not all samples could be processed without some delay and instead they were stored at 4°C. Isolation followed as soon as possible, with a preference of processing the biofilm samples before the effluent. R2A plates were left to grow at room temperature for 5 days, at which point plate counts were obtained and total DNA isolation was performed as outlined below.

DNA Isolation

Total DNA isolation was performed using two different techniques depending on the type of sample being processed. For both biofilm and effluent samples from the annular reactors, the MoBio Soil DNA isolation kit was used as follows. The total sample was centrifuged at 20,000 x *g* for 30 minutes at 4°C. Most of the supernatant was removed using a 25 mL serological pipette, leaving behind 2 mL of the supernatant. The resulting pellet was resuspended in the 2 mL of supernatant by vortex, transferred into a 2 mL screw cap micro tube, and centrifuged at 20,000 x *g* for 20 minutes at room temperature. After removing the supernatant the contents of one of the bead solution tubes (from the MoBio kit) was added to the pellet. The bead solution contains a buffer and it begins to dissolve humic acids. After vortexing to resuspend the pellet, 60 µL of Solution S1 was added, and the tube was inverted several times. Solution S1 contains sodium dodecyl sulphate (SDS), which helps in cell lysis. Two hundred microlitres of solution IRS (inhibitor removal solution) was added and the tube was inverted several times. Solution IRS is a proprietary reagent designed to precipitate humic acids and other PCR inhibitors. Bead beating was then carried out by taping the tubes to a flat vortex and vortexing at maximum speed for 10 minutes. The tubes were then centrifuged at 10,000 x *g* for 30 seconds and the supernatant was transferred to a new 1.5 mL microcentrifuge tube. Two hundred and fifty microlitres of solution S2 was then added to each tube, vortexed for 5 seconds, and incubated at 4°C for 5 minutes. Solution S2 is a proprietary reagent that contains a protein precipitation compound. The precipitated proteins were then pelleted by centrifuging the tubes at 10,000 x *g* for 1 minute at room temperature. The supernatant was transferred to a clean 2 mL microcentrifuge tube, 1.3 mL of solution S3 was added, and vortexed for 5 seconds. Solution S3 is a DNA-binding salt solution. The solution was then passed through a spin filter by a series of centrifugation steps at 10,000 x *g* for 1 minute each, in order to bind the DNA to the silica filter. The spin filter was then washed by applying 300 µL of Solution S4 (ethanol based wash solution used to remove residues of salt, humic acid and other contaminants) and centrifuging for 30 seconds at 10,000 x *g*. Any residual solution S4 was removed from the filter with a subsequent centrifugation at 10,000 x *g* for 1 minute, before transferring the spin filter column to a clean 1.5 mL microcentrifuge tube. The DNA was eluted from the spin filter column by adding 50 µL of solution S5 (10mM Tris pH 8) and letting it sit for 1 minute before centrifugation at 10,000 x *g* for 30 seconds. The isolated DNA was then stored at -20°C until used.

For extracting DNA from heterotrophs, total bacteria were harvested from the plates by using a sterile glass rod with rubber spatula to scrape all of the colonies off the plate and transfer them into a 25 mL screw cap tube containing 5 mL of 0.85 % saline. After vortexing, 1 mL was transferred to a 2 mL screw cap tube and centrifuged at 5,000 x *g* for 10 minutes. Total DNA was isolated from the resulting pellet using the Qiagen DNeasy Tissue kit as follows. The bacterial pellet was resuspended in 100 μl of lysis buffer (20 mM Tris·Cl, 2 mM sodium EDTA, 1.2 % Triton® X-100, 20 mg/mL lysozyme) and incubated at 37°C for 30 minutes. Two hundred microlitres of buffer AL and 25 μl of proteinase K were then added to each tube, mixed by vortexing, and then incubated at 95°C for 20 minutes. Two hundred microlitres of 100 % ethanol was then added before transferring the solution into a DNeasy mini spin column and centrifuging at 6,000 x *g* for 1 minute. After discarding the flow-through, 500 μL of buffer AW1 was added to the spin column and then centrifuged for 1 minute at 6,000 x *g*. The column was washed again with 500 μL of buffer AW2 by centrifuging for 3 minutes at 20,000 x *g*, and then placed in a clean 1.5 mL microcentrifuge tube. To elute the DNA from the column 150 μl of buffer AE was applied to the column and let sit for 1 minute at room temperature before centrifuging for 1 minute at 6,000 x *g*. The isolated DNA was then stored at -20°C until used.

DNA Quantification

DNA was quantified using a Beckman DU® 640B spectrophotometer so that approximately the same quantity could be used for each sample. The DNA sample was diluted in ultra pure molecular biology grade water by pipetting 5 μl of sample into 95 μl of water in a 0.5 mL PCR tube. The diluted sample was read three times in the spectrophotometer at three wavelengths, 260 nm 280 nm and 320 nm, against a blank consisting of 5 μl of the same elution buffer used in the DNA isolation as the sample and 95 μl of ultra pure water. The absorbance values obtained at 260 nm were then averaged and used in the following calculation in order to obtain the DNA concentration of the original sample.

Concentration = Absorbance x Extinction Coefficient x Dilution Factor x Pathlength

Where: Dilution Factor = Final Volume / Initial Volume of Sample

The concentration is obtained in μg/mL, the absorbance is the average of the reading at 260 nm, the extinction coefficient is 50 μg/mL, the dilution factor is 20 (100 μl / 5 μl), and the path length is 1cm.

$$\therefore \text{[DNA (μg/mL)]} = Abs_{260} \times 50 \text{ μg/mL} \times 20 \times 1\text{cm}$$

PCR Amplification for DGGE

The target amplified in the PCR reaction was the variable V3 region in the 16S rRNA gene consisting of approximately 193 base pairs (positions 341 to 534 in the *E. coli* 16S rRNA gene) Each PCR reaction was made to a final concentration of 1X *Taq* buffer with KCl (Fermentas), 1.5 mM $MgCl_2$, 200 μM dNTP's (Amersham Biosciences), 2 % DMSO, 0.16 μM of each 341F-GC (contains a 40 bp 5' G-C clamp) and 518R primers, and 2.5 units of *Taq* polymerase (Fermentas), with a final total volume of 25 μl in 0.5 mL thin-walled PCR tubes (Muyzer et al 1993, 1998; Muyzer and Smalla 1998, Muyzer 1999). A total of 30 ng of DNA template prepared by the above methods was used for each sample. A Thermolyne Amplitron® II thermocycler was used with the following PCR cycle conditions: initial denaturation at 94°C for 2 minutes, 35 cycles of 94°C for 30 seconds, 60°C for 45 seconds, and 72°C for 45 seconds, after which a final extension step of 5 minutes at 72°C was added, followed by a hold at 4°C. A negative control was also run using ultra pure dH_2O in place of template. Following amplification 8 µl of 6X loading dye (Fermentas) was added to each PCR reaction.

Denaturing Gradient Gel Electrophoresis

Denaturing gradient gel electrophoresis (DGGE) separates DNA fragments of the same size on the basis of sequence under a denaturing gradient. DGGE was performed using the Dcode® Universal Mutation Detection System (Bio-Rad Laboratories). Twelve microlitres of the PCR amplified DNA was analyzed on a 8% polyacrylamide gel consisting of a 40-60% gradient of denaturant (urea and formamide). A small amount of silicon grease was applied to the sides of the glass plates containing the spacers in order to prevent 'smiling' of the bands on the edges of the gel. All gels were run in 0.5 X TAE at a constant voltage of 37 V for 18 hours at 65 °C. An initial 200 V for 2 minutes was used to quickly draw the DNA into the gel. Bands were visualized by silver stain using the method below. To ensure even running of the gels and to compare bands from different gels a DGGE marker, prepared in house from commonly found sequences, was loaded in the first and last lanes of each gel.

Silver Staining and Gel Drying

Silver staining was used in preference to ethidium bromide so that a copy of the gel can be retained and so that bands can be easily cut out. Silver staining was carried out directly after DGGE in a Pyrex® casserole dish. Fixation was carried out in a 300ml solution of 40 % methanol and 10 % acetic acid for 1 hour on a Labline Maxi Rotator at medium speed. This was followed by three washes in 30 % ethanol (approximately 330 mL) for 10 minutes each. Reduction was then carried out in 300 mL of 0.02 % sodium thiosulfate for 1 minute while shaking by hand. This was followed by three washes in deionized H_2O for 20 seconds each while shaking by hand. The gels were then incubated in 300 mL of 0.2 % silver nitrate solution for 20 minutes while shaking on a rotator at medium speed. Three washes using deionized H_2O for 20 seconds each while shaking by hand was then applied to remove the excess silver nitrate from the gels. Developing was then effected by incubating the gels in a 500 mL solution of 3 % sodium carbonate and 0.05 % formaldehyde for 3-5 minutes or until desired colour was reached. There were three final washes in deionized water for 20 seconds each before incubating the gels for 5 minutes in 300 mL of stop reagent consisting of 0.5% glycine. In order to prevent cracking of the gels during drying, they were incubated in a final glycerol solution consisting of 20 % ethanol and 10 % glycerol for 1 hour. Gel drying was obtained by sandwiching the gel in two pieces of cellophane using a Tut's Tomb (Research Products International) gel drying frame, and leaving it out at room temperature overnight until dry. Excess cellophane was then trimmed from the gels, and they were labelled and stored in a three-ring binder. Each gel was also electronically scanned using a Hewlett-Packard Scanjet IIc scanner at 300 dpi, scanning both the fully labelled insert and only the gel itself.

Excising Bands from DGGE Gels

Bands were cut from the gels by using techniques to avoid DNA cross contamination. Either a new scalpel was used for each band, or the scalpel was very well flamed between bands.

Only half of the band at most was cut from the gel, making sure to cut only the band by cutting into it rather then around it. The excised band was then put into a clean microcentrifuge tube containing 200μl of ultra pure molecular biology grade H_2O and vortexed for 2 minutes. The two pieces of cellophane and all of the H_2O were then removed leaving behind only the excised band in the tube, and another 200 μl of ultra pure H_2O was added to wash the band once again by vortexing. This was repeated at least two more times. After removing the H_2O for the last time, 75 μl of sigma H_2O was added to the tube and the band was squished into small pieces using a pipette tip and the side of the tube. The tubes were then placed in a standard heatblock (VWR Scientific) and heated to 95°C for 20 minutes, with a brief vortex at the 10 minute mark. The tubes were then incubated at 4°C overnight. The sample was then ready to use and was stored at -20°C.

Reamplification for Cloning

Reamplification used the same primers but without the GC clamp. PCR was similar to that described above with a final concentration of 1X *Taq* buffer with KCl (Fermentas), 1.5 mM $MgCl_2$, 200 μM dNTP's (Amersham Biosciences), 2 % DMSO, 0.16 μM of each 341F and 518R primers, and 2.5 units of *Taq* polymerase (Fermentas), with a final total volume of 25 μl in 0.5 mL thin-walled PCR tubes. A total of 15.2 μl of the DNA template from the excised band prepared by the above method was used for each sample. A Thermolyne Amplitron® II thermocycler was used with the following PCR cycle conditions: initial denaturation at 94°C for 2 minutes, 35 cycles of 94°C for 30 seconds, 57°C for 45 seconds, and 72°C for 45 seconds, after which a final extension step of 5 minutes at 72°C was added, followed by a hold at 4°C. A negative control was also run using ultra pure dH_2O in place of template. Following amplification 5 μl of 6X loading dye (Fermentas) was added to each tube, and the amplicons were run on a 0.8 % low melting point agarose gel against a 100 bp DNA ladder (GeneRuler™, Fermentas). The single band in approximately the 200 bp region was cut out using sterile microbiological techniques, and placed in a clean microcentrifuge tube for use in the following cloning technique.

Cloning and Plasmid Isolation

Cloning was done to ensure that only one copy of the gene was present and sequencing is possible. It was achieved using Invitrogen's TOPO TA Cloning kit with pCR 2.1-TOPO vector and One Shot TOP10 Chemically Competent *E. coli*. The manufacturer instructions were modified as follows. A round bottom 96 well plate was used for all reactions. As stated above, the cut out band from the low melting point agarose gel is used as template for the cloning procedure. We used 4μl of the melted template sample in our reactions with the vector, and it was then incubated for 10 minutes at 37°C. Only 25 μl of the One Shot *E. coli* was applied to the entire vector reaction mixture, and it was then incubated on ice for 30 minutes. Heat shock was proceeded as outlined by manufacturer (42°C for 40 seconds), after which it was placed back on ice and 125 μl of SOC medium was added. The mixture was then put on an orbital shaker (Forma Scientific) at 200 rpm for 1 hour at 37°C. The entire mixture was spread on a pre warmed Luria-Bertani (LB) plate containing 100 μg/mL ampicillin, 50 μg/mL IPTG, and 30 μg/mL X-gal, and incubated at 37°C overnight. Three white colonies from each plate were individually picked using Calibrated Disposable Inoculating Loops (BD Biosciences) and were transferred to a conical screw cap tube (Sarstedt) containing 3 mL of LB broth containing 100 μg/mL ampicillin, and were incubated overnight at 37°C on an orbital shaker at 200 rpm with the caps loosely fastened. The suspension was then taken through the plasmid isolation step by centrifuging 1 mL at 7000 x g for 10 minutes, and using the pellet in the QIAprep Miniprep kit (Qiagen) with the manufacture directions. DNA quantification of the plasmid isolations was then preformed as above, and plasmid isolations were kept at -20°C until further use.

Plasmid Insert Checking and Identification

The plasmid inserts were crosschecked against the original sample the band was excised from by PCR amplifying out the plasmid insert as above using DGGE primers and PCR program. For template 0.5 μl of the plasmid isolation was used. The amplicons were then run on a 40-60 % DGGE gel (as above), along side the original sample the band of interest was excised from, and containing the DGGE marker described above on each end of the gel. If at least one of the plasmid isolation bands matches the migration of the band of interest, and it is a single band clear of spurious by-products, then that sample will be identified. If none of the plasmid isolation bands match the position then 3 more colonies from the cloning plate for that sample will be picked and isolated, until such point that the correct migration is obtained. Identification was obtained by directly sequencing the plasmid insert using the M13f primer. Sequencing was performed by the University of Ottawa Core DNA Sequencing and Synthesis Facility. Obtained sequences were then modified by removing the extra plasmid sequence data on either end of the insert, and put into FASTA format containing the sample number as the identifier. The modified sequences were then aligned with both NCBI's BLASTn and the Ribosomal Database Project's Sequence Match online database programs for identification.

Acquisition of Antibiotic Resistance in DWDS Biofilms

Even if antibiotic resistance of strains in the source water is affected by exposure to the disinfectant, it would not be easy to demonstrate. We therefore used *Escherichia coli* as a model organism to explore possible links between sublethal disinfectant exposure and antibiotic resistance. The wild type strain used was one with known genetic sequence and an isogenic mismatch repair mutant was also used to represent mutator strains; further details are given below.

Media and Antibiotics

Mueller Hinton medium and m-FC agar medium were purchased from Oxoid Canada Inc. and Difco Laboratories., respectively. Luria Bertani (LB) broth was prepared in house from its constituents. Membrane microarray starter kit, and Multi-Spense for 384-well plates were obtained from V&P Scientific Inc. Rifampicin, Tetracyclin, Chloramphenicol, Gentamicin, Ampicillin, Nalidixic, Sulfisoxazol, Trimethoprim and Cefoperazone were purchased from Sigma. BCIG (*5-bromo-6-chloro-3-indolyl-β-D-glucuronide*) was purchased form Diagnostic Chemicals Limited. Single well and 384-well Omni trays, nitrocellulose membrane and all other plastic ware were purchased from VWR Canada Inc.

Bacterial strains

E.coli K12 (MG1655) and *E.coli* TX3496 (*MG1655 mutS::Ω [Km^r; EcoRV]*) , a *mutS* mutant strain that was derived from wild type MG1655 by P_1 transduction i.e. *MG1655xP$_1$virTX3451*, were obtained from Dr. Winkler M.E. (Department of Microbiology and Molecular Genetics, University of Texas Houston Medical School, 77030-1501, USA). The *E.coli* ATCC 25922 strain, recommended by NCCLS, was used as an external control in antibiotic susceptibility tests and during screening for resistant mutants. The internal control was *E. coli* K12 (MG1655) that had not been exposed in DWDS biofilms.

Growth of biofilms and isolation of antibiotic resistant mutants of E. coli

Ideally, biofilms grown without and with two levels of disinfectant should have been run in parallel. Unfortunately, due to the large numbers of variables and the practical constraints to the system it was necessary to run these treatments in series. Prior to introduction of *E.coli* strains to the annular reactors natural biofilms of resident heterotrophic bacteria were formed by running the tap water provided by Halifax Regional Water Commission through granular activated carbon for dechlorination. *E.coli* K12 and *E.coli mutS* were introduced and established in biofilms by inoculating these strains into odd numbered and even numbered reactors (AR1-AR8), respectively. After 2-3 weeks of acclimation without treatment, the reactors were treated sequentially, first with a low level of disinfectant (~0.2 µg/mL) and, after sampling from the reactors for 2-3 weeks, then with a higher level of disinfectant (~1 mg/L) for a further 2-3 weeks. These experiments were run with chlorine, monochloramine and chlorine dioxide and each disinfectant was tested with and without the addition of upstream UV for biofilms grown on two different substrates (cast iron and polycarbonate). Antibiotic resistant biofilm *E.coli* isolates derived from the wild type *E.coli* K12 and its otherwise isogenic *mutS* mutant strain were screened in three steps.

In the first step, done in Halifax, enteric coliform bacteria in the mixed bacterial population from each coupon were plated on coliform selective m-Endo plates and shipped to Ottawa. In Ottawa, the colonies from m-Endo plates that showed characteristic metallic sheen were picked and grown, along with proper controls, in 384-well Omni tray plates containing 50 µl LB broth medium per well at 37°C overnight. A working replica of each 384-well plate was prepared using 384 Pin MULTI-BLOT™ Replicator (V&P Scientific Inc.), and both sets of plates were frozen in -80°C freezer after addition of 50 µl sterile 20% glycerol. The A_{600} values for a total of 95 random samples from each of the working 384-well plates were measured to ensure that the culture densities among different isolates did not show any significant variation, which correlated to 10^4-10^5 cfu/ul.

In the second screening step, the isolates grown in 384-well plates were replicated on nitrocellulose membrane that was overlaid on single well Omni trays containing m-FC agar medium supplemented with 100 µg/mL BCIG. On m-FC/BCIG agar medium the glucuronidase positive *E.coli* appear as blue colonies that was considered as the starting population of *E.coli* isolates for antibiotic resistance screening.

In the third step of screening the isolates were replicated onto Muller Hinton agar plates that were supplemented with Rifampicin (40 µg/mL), Tetracycline (3-6.25 µg/mL), Ampicillin (32 µg/mL), Chloramphenicol (16 µg/mL), Nalidixic acid (8 µg/mL), Trimethoprim (3 µg/mL), Sulfisoxazole (64 µg/mL), Cefoperazone (3 µg/mL) and Gentamicin (4 µg/mL). These antibiotic concentrations were chosen to maintain adequate selective pressure, i.e. 2 to 4-fold higher than measured MIC values for controls strains, on isolates that would prevent the growth of the positive controls over 18-24 hours of incubation at 37°C.

Dilution MIC assay

In order to maintain an appropriate level of antibiotic selective pressure during screening for antibiotic resistant isolates, the MIC values of control *E.coli* strains against the selected antibiotics were determined using agar dilution MIC assay. Half dilution series of the antibiotics (0.0625 to128 μg/mL) were prepared in Mueller Hinton broth using 96-well plates. Overnight cultures of *E.coli* K12, *E.coli* TX3496 and *E.coli* ATCC 25922 were pelleted and washed in sterile saline. The pellets were resuspended in appropriate volume of saline to achieve A_{600} of 0.05 that corresponded to about 10^7 cells/mL in total volume of 20-30 mL. The prepared suspensions were transferred into a three single well Omni trays. The trays were used to inoculate about 1 μl samples (~10^4 cells) into each well of the 96-well plates that contained the antibiotic dilution series. Plates were incubated at 37°C for 18-24 hours. The visible appearance of bacterial growth in the wells was used to determine the MIC values.

PCR and cloning

rpoB specific pairs of primers were designed (Figure 3.7) considering the *E.coli* whole genome as template to amplify the cluster II region of *rpoB* gene using Oligo primer design program (Molecular Biology Insights, Inc.). Standard PCR reactions were carried out in a total volume of 50 μl using 0.05 μg of pure genomic DNA and 2.5 units of Taq polymerase (Invitrogen Life Technologies). The thermal cycling conditions were optimized to produce a single PCR product band of 636 base pairs using the following conditions: Pre-dwell: 94°C/5 min, 25 cycles of denaturation at 94°C/1 min, annealing at 48°C/30 sec, extension at 72°C/30 sec followed by one cycle at 72°C/30 min.

The amplified product that contained the cluster II region of the *rpoB* gene was further verified by restriction analysis using BclI digestion and cloned into pCR4-TOPO vector (Molecular Biology Insights, Inc.) and screened on LB agar plates containing X-gal and ampicillin. The sequence of the cloned fragments from the control strains were verified before applying the same PCR/cloning protocol to the candidate isolates.

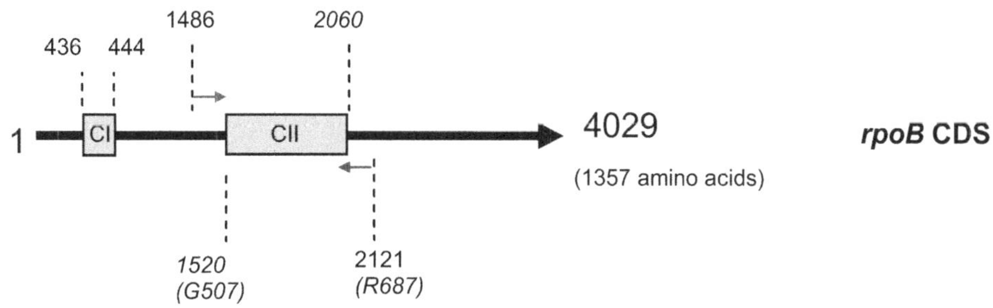

Product size: 636 bp

→ Ec rpoB+ 5' AAG CCG ATT TCC GCA GCA GTG AA 3' (23 mer)

← Ec rpoB- 5' AGC ACG TTC CAT ACC AGT ACC AA 3' (23 mer)

Figure 3.7 Design for *rpoB* primers.

Database management

A total of 12 databases that include information pertaining to the characteristics of each single isolate were developed and maintained in Excel format. Appropriate functions were incorporated to the datasheets in order to generate output summary tables to illustrate all the characterization information available for each isolate including their multiple resistance characteristics. Databases were further used to categorize the isolates on the basis of their overall phenotypic behaviours and to sample representative isolates from each category for further molecular analyses.

Statistical Analyses

Statistical tests were repeated for the various combinations of disinfectants type. In addition, statistical tests compared the significant differences between the average influent and effluent values for the water quality parameters measured. The level of significance that was used for all tests was $\alpha = 0.05$. Statistical procedures followed were an analysis of variance (ANOVA) test, as described by Box et al (1978).

CHAPTER 4
LABORATORY SCALE DISINFECTION STUDIES

INTRODUCTION

Coliforms bacteria are used worldwide as indicators of microbiological quality and safety of drinking water (Besner *et al,* 2002). Total coliforms and *Escherichia coli* (*E. coli*) are routinely monitored at public utilities and their detection is often an indication of a) inadequate treatment, b) breech in distribution system integrity and/ or c) loss of disinfectant residual (Williams *et al,* 2003; Craun & Calderon, 2001; LeChevallier, 1990).

Once introduced into distribution systems, the presence of *E. coli* and total coliforms become a great public health concern. Several outbreaks worldwide have been associated with the presence of *E. coli* O157 in drinking water supplies such as: Walkerton, ON, Canada, Highland Region, Scotland and Cabool, MO, USA. These drinking water outbreaks resulted in people becoming severely ill and in some cases even dying. Numerous peer reviewed studies have shown that coliforms are capable of persisting in drinking water distribution system biofilms in the presence of disinfectants (Camper *et al*, 1999; Camper *et al*, 1991; Williams and Braun-Howland, 2003; LeChevallier, 1990, Norton and LeChevallier, 1997.) Consequently, the ability of coliforms to persist in distribution system biofilms is of significance to the water industry. It is an issue that requires more investigation and understanding before researchers can develop appropriate solutions and prevention measures.

This chapter examined the effects of pipe material, chemical disinfectant and UV disinfection on two strains of *E. coli K12:* a wild strain and a derived mutant. In addition, phenotypical changes of *E. coli* under these conditions will be discussed.

RESULTS

Disinfectant Residuals

The target disinfectant residuals for each disinfectant trial are presented in Table 4.1. These residuals are based on the disinfectant concentration in the effluent of the ARs. The average actual residuals achieved during the trials are presented in Tables 4.2 - 4.4. The tables are grouped according to chemical disinfectant and residual (low, high). Since the disinfectant residuals vary from low to high disinfection, inactivation data was conducted on the basis of average disinfectant residual and contact time (Ct). Contact time was calculated by using the measured residual (C) and the hydraulic retention time of the ARs (t) using Equation 4.1. The hydraulic retention time of the ARs throughout the experiments was 6 hours.

Ct= Average Disinfectant Residual x HRT (4.1)
Where HRT= 6 hour

Table 4.1
Target Disinfectant Residuals

Disinfectant	Target low residual (mg/L)	Target high residual (mg/L)
Chlorine	0.2	1.0
Chlorine Dioxide	0.2	1.0
Monochloramine	1.0	2.0

From the high standard deviations in Tables 4.2 to 4.4 it is evident that it was difficult to maintain disinfectant residuals in the ARs. It was much more difficult to maintain a disinfectant residual and meet the target residual in the cast iron ARs than in the polycarbonate ARs (Figures 4.1–4.2). The increased disinfectant demand due to iron oxidation made it difficult to sustain a consistent residual in the cast iron annular reactors for all three chemical disinfectants.

Table 4.2
Disinfectant residuals for chlorine trial

Treatment	Low Cl_2 disinfection			High Cl_2 disinfection		
	Target residual (mg/L)	Average measured residual (mg/L) (Standard deviation)	Ct (mg/L•min)	Target residual (mg/L)	Average measured residual (mg/L) (Standard deviation)	Ct (mg/L•min)
Pre-UV-CI - K12	0.2	0.27 (0.29)	97.2	1.0	0.45 (0.61)	162
Pre-UV-CI-Mutant	0.2	0.15 (0.17)	54	1.0	0.32 (0.52)	115.2
Pre-UV-PC-K12	0.2	0.26 (0.33)	93.6	1.0	0.70 (0.47)	252
Pre-UV-PC-Mutant	0.2	0.37 (0.52)	133.2	1.0	0.91 (0.86)	327.6
CI - K12	0.2	0.18 (0.22)	64.8	1.0	0.41 (0.54)	147.6
CI-Mutant	0.2	0.11 (0.10)	39.6	1.0	0.48 (0.69)	172.8
PC-K12	0.2	0.43 (0.48)	154.8	1.0	0.94 (0.44)	338.4
PC-Mutant	0.2	0.32 (0.32)	115.2	1.0	1.07 (0.37)	385.2
Post-UV-CI - K12	0.2	0.08 (0.18)	28.8	1.0	0.72 (0.75)	259.2
Post-UV-CI-Mutant	0.2	0.24 (0.43)	86.4	1.0	0.34 (0.54)	122.4
Post-UV-PC-K12	0.2	0.32 (0.40)	115.2	1.0	0.6 (0.40)	216
Post-UV-PC-Mutant	0.2	0.25 (0.27)	90	1.0	0.87 (0.61)	313.2

Table 4.3
Disinfectant residuals for chlorine dioxide trial

Treatment	Low ClO_2 disinfection			High ClO_2 disinfection		
	Target residual (mg/L)	Average measured residual (mg/L) (Standard deviation)	Ct (mg/L•min)	Target residual (mg/L)	Average measured residual (mg/L) (Standard deviation)	Ct (mg/L•min)
Pre-UV-CI - K12	0.2	0.33 (0.44)	118.8	1.0	0.79 (0.44)	284.4
Pre-UV-CI-Mutant	0.2	0.23 (0.29)	82.8	1.0	0.94 (0.52)	338.4
Pre-UV-PC-K12	0.2	0.26 (0.18)	93.6	1.0	0.88 (0.43)	316.8
Pre-UV-PC-Mutant	0.2	0.26 (0.17)	93.6	1.0	0.9 (0.35)	324
CI - K12	0.2	0.14 (0.16)	50.4	1.0	1.18 (0.56)	424.8
CI-Mutant	0.2	0.19 (0.16)	68.4	1.0	0.35 (0.23)	126
PC-K12	0.2	0.16 (0.13)	57.6	1.0	1.06 (0.4)	381.6
PC-Mutant	0.2	0.18 (0.13)	64.8	1.0	0.9 (0.33)	324
Post-UV-CI - K12	0.2	0.26 (0.23)	93.4	1.0	0.47 (0.55)	169.2
Post-UV-PC-Mutant	0.2	0.17 (0.14)	61.2	1.0	0.89 (0.34)	320.4

Table 4.4
Disinfectant residuals for chloramine trial

Treatment	Low NH_2Cl disinfection			High NH_2Cl disinfection		
	Target residual (mg/L)	Average measured residual (mg/L) (Standard deviation)	Ct (mg/L•min)	Target residual (mg/L)	Average measured residual (mg/L) (Standard deviation)	Ct (mg/L•min)
Pre-UV-CI - K12	1.0	0.14 (0.14)	50.4	2.0	1.07 (1.22)	385.2
Pre-UV-CI-Mutant	1.0	0.16 (0.20)	57.6	2.0	0.99 (1.05)	356.4
Pre-UV-PC-K12	1.0	0.77 (0.40)	277.2	2.0	1.65 (0.68)	594
Pre-UV-PC-Mutant	1.0	0.85 (0.48)	306	2.0	1.47 (0.8)	529.2
CI - K12	1.0	0.30 (0.31)	108	2.0	1.52 (1.45)	547.2
CI-Mutant	1.0	0.30 (0.37)	108	2.0	1.21 (1.32)	435.6
PC-K12	1.0	0.71 (0.44)	255.6	2.0	1.65 (0.85)	594
PC-Mutant	1.0	0.79 (0.43)	284.4	2.0	1.45 (0.74)	522
Post-UV-CI - K12	1.0	0.28 (0.37)	100.8	2.0	1.36 (1.39)	489.6
Post-UV-PC-Mutant	1.0	0.60 (0.34)	216	2.0	1.23 (0.56)	442.8

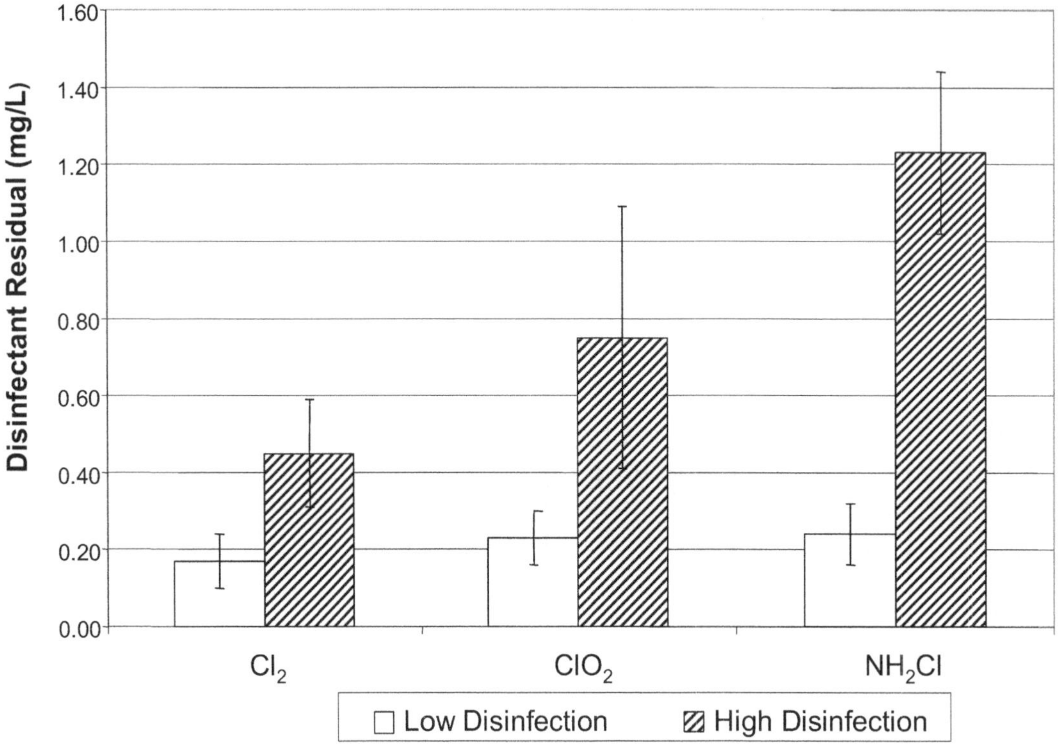

Figure 4.1 Average disinfectant residuals in cast iron ARs

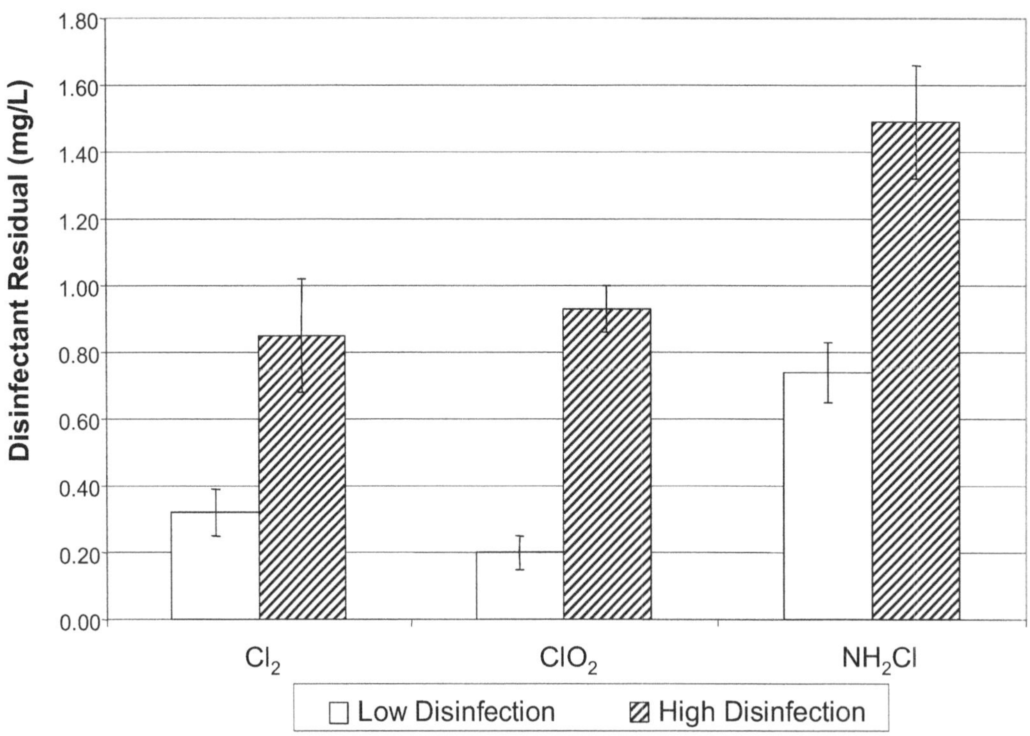

Figure 4.2 Average disinfectant residuals in polycarbonate ARs

Effluent Heterotrophic Bacteria

The baseline HPCs into the annular reactors ranged from 5.0 x 10^1 CFU/ mL to 7.0 x 10^3 CFU/mL for both UV and Non-UV treated systems. Effluent heterotrophic bacteria log reductions for low and high disinfection are presented in Table 4.5 for Cl_2, ClO_2 and NH_2Cl, respectively. Log reductions were calculated using average HPC data from the acclimation, low disinfection and high disinfection phases. Equation 4.2 was used to calculate log reduction.

$$\text{Log Reduction} = -\log \{\text{Avg. HPC}_{\text{disinfection}} / \text{Avg. HPC}_{\text{acclimation}}\} \quad (4.2)$$

There were no statistically significant differences in effluent and biofilm HPCs between ARs spiked with *E. coli* K12 and those spiked with the mutant strain (p= 0.054- 0.425). All discussion, figures and tables from this point on refer to ARs that were spiked with *E. coli* K12 unless identified otherwise.

Table 4.5
Effluent HPC log reductions

Treatment	*Effluent* HPCs- Log reduction during disinfection					
	Low Cl_2	High Cl_2	Low ClO_2	High ClO_2	Low NH_2Cl	High NH_2Cl
Pre-UV-CI - K12	2.08	3.00	0.80	2.88	0.08	2.68
Pre-UV-CI-Mutant	1.52	1.94	1.09	4.19	0.10	2.63
Pre-UV-PC-K12	2.68	4.10	3.24	3.61	1.23	2.13
Pre-UV-PC-Mutant	3.97	3.88	2.69	4.01	2.09	2.85
CI - K12	2.09	2.95	1.77	2.88	1.47	2.74
CI-Mutant	2.01	3.96	1.37	2.59	0.74	2.64
PC-K12	2.93	4.08	3.09	4.39	1.32	1.36
PC-Mutant	2.01	3.47	2.57	3.97	1.60	1.57
Post-UV-CI - K12	1.64	2.90	1.33	4.48	0.33	2.42
Post-UV-CI-Mutant	1.97	2.47	-	-	-	-
Post-UV-PC-K12	2.58	2.96	2.53	4.54	1.89	2.79
Post UV PC Mutant	3.50	3.97		-	-	-

Effect of UV Disinfection

As mentioned in Chapter 3, three UV treatments were explored throughout the different disinfectant trials (Cl_2, ClO_2 & NH_2Cl). Pre-UV treatment refers to when UV disinfection occurred prior to the *E. coli* spikes and Post-UV treatment occurred after the *E. coli* spikes. In addition, four ARs in each run received no UV treatment.

Figures 4.3- 4.8 demonstrate the differences in effluent heterotrophic bacteria between Pre-UV, Non-UV and Post-UV for cast iron and polycarbonate ARs for each chemical disinfectant. In each figure, the corresponding Ct values during low and high disinfection are displayed.

Chlorine Trial

In general, log removals for suspended HPCs in cast iron ARs during the chlorine trial were similar between ARs treated with UV (either Pre or Post) and those not treated with UV (Table 4.5 and Figure 4.3). In Figure 4.3, the Non- UV treated AR achieved higher log removals than the Post-UV AR during low disinfection (Table 4.5). Interestingly, the Post-UV treated AR and the Non-UV treated AR had log removals during low disinfection of 1.64 and 2.09 log, respectively. This difference could be attributed to the lower Ct maintained in the Post-UV AR. It is anticipated that had the Ct value been comparable equal or better log removal would have been achieved. Similar trends were observed for polycarbonate ARs where Pre-UV and Post-UV treatment achieved comparable log removals to Non-UV despite their lower Ct values (Figure 4.4).

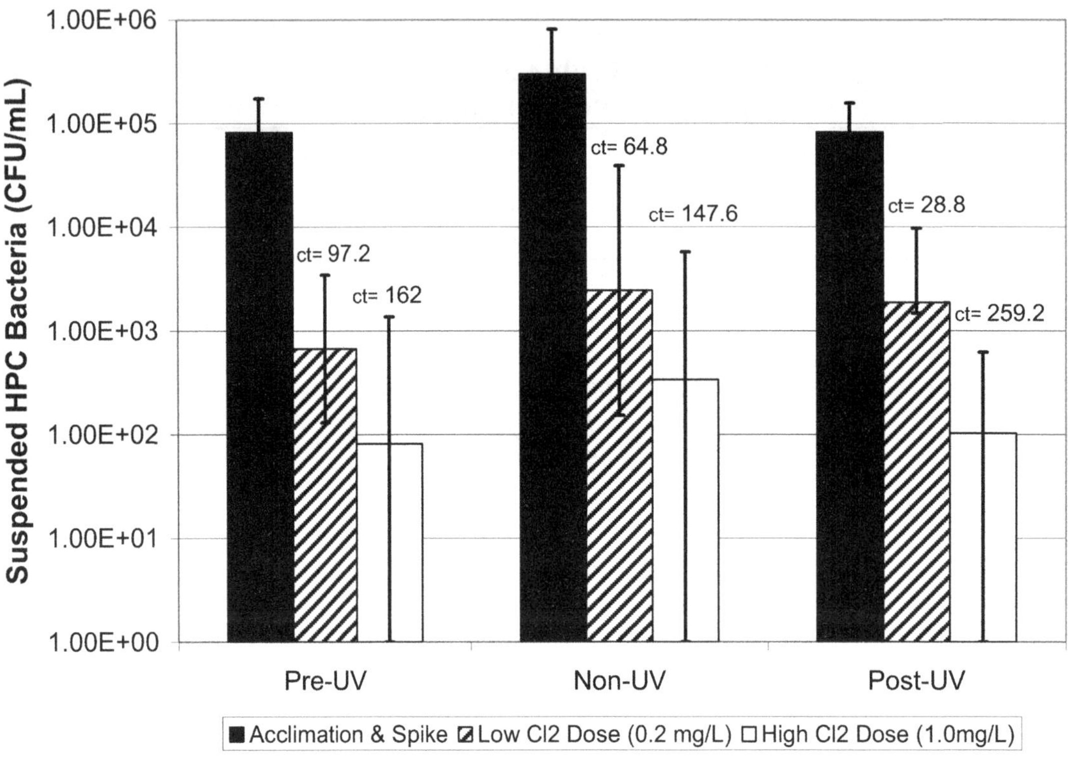

Figure 4.3 Suspended heterotrophic bacteria in cast iron ARs during chlorine disinfection

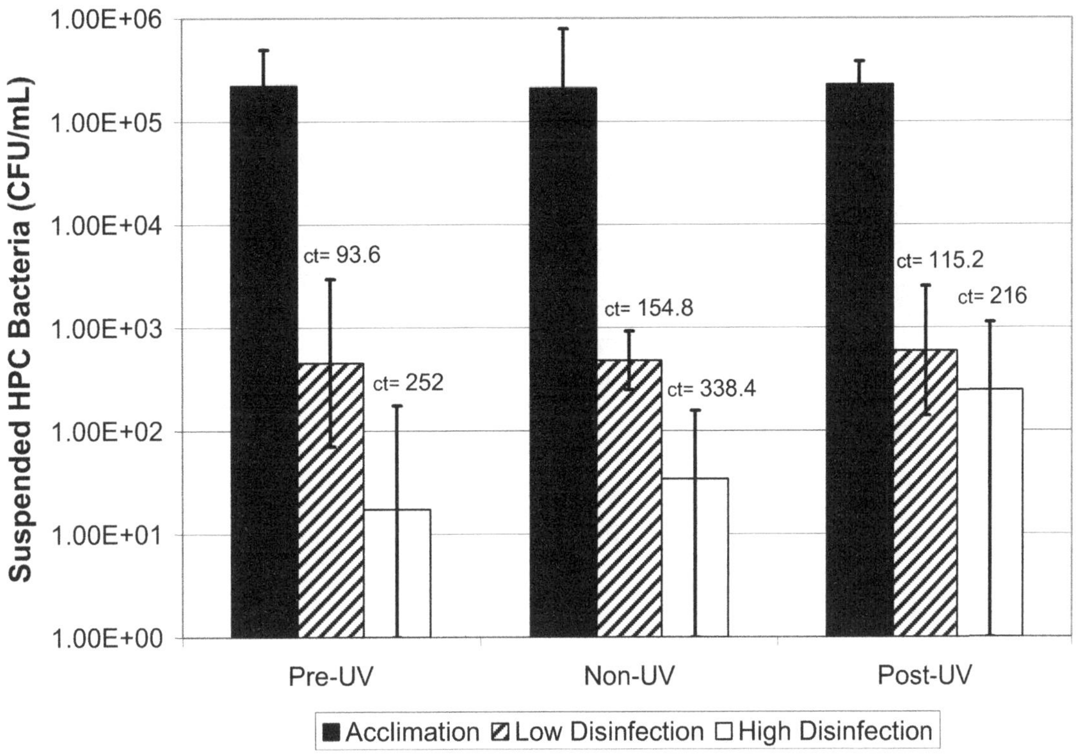

Figure 4.4 Suspended heterotrophic bacteria in polycarbonate ARs during chlorine disinfection

However, according to statistical analysis there were no significant differences between Pre-UV, Non-UV and Post-UV in both cast iron and polycarbonate reactors. The large error bars in Figures 4.3 and 4.4 were attributed to changes in disinfectant residual during disinfectant ramping and overall maintenance of the residuals.

Chlorine Dioxide Trial

Log reductions in effluent HPCs were comparable in cast iron ARs during disinfection; they ranged from 0.79- 1.38 log. Strangely, the AR that received no-UV treatment had the lowest Ct value and yet had the highest log removal. This could be because the iron in the Pre and Post UV ARs was more easily oxidized during this period. Higher average turbidity values during disinfection of 4.96 NTU (Pre-UV) and 10.11 NTU (Post-UV) versus 2.69 NTU (Non-UV) indicate that is may be the case. During high disinfection the opposite occurred, a larger Ct was maintained in the Non-UV AR than the Pre and Post UV ARs and yet achieved the lowest log removal of 2.12 log (Figure 4.5 and Table 4.5).

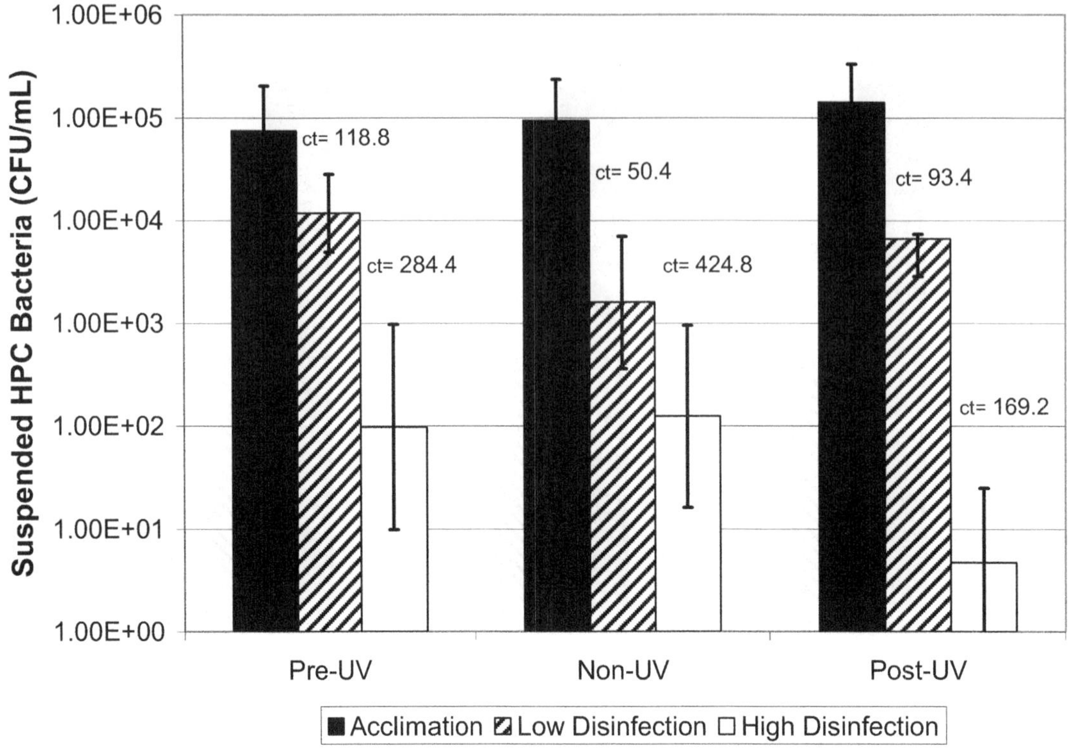

Figure 4.5 Suspended heterotrophic bacteria in cast iron ARs during chlorine dioxide disinfection

The Post-UV cast iron AR had the highest log reduction of 4.71 log during this period. Although it seems as though the UV reactors had the greatest log removals with respect to disinfectant residual, these results were not shown to be significant.

Polycarbonate ARs showed similar log reductions during low (2.3-2.78 log) and high (3.44- 3.90 log) disinfection (Figure 4.6 and Table 4.5). However, during high disinfection, the disinfectant residuals maintained in the Pre-UV (Ct= 338.4) and Post-UV (Ct= 320.4) ARs were lower than that of the Non-UV (Ct= 381.6) AR. This suggests that had all the Ct values been equivalent, the UV treated reactors would have shown higher log removals than the Non-UV reactor. According to statistically analysis there is no significant difference between the three treatments on effluent HPCs in polycarbonate ARs.

The error bars for Figures 4.5 and 4.6 are much smaller than those in Figures 4.2 and 4.3, indicating that chlorine dioxide residuals were much more constant and easier to maintain than chlorine residuals.

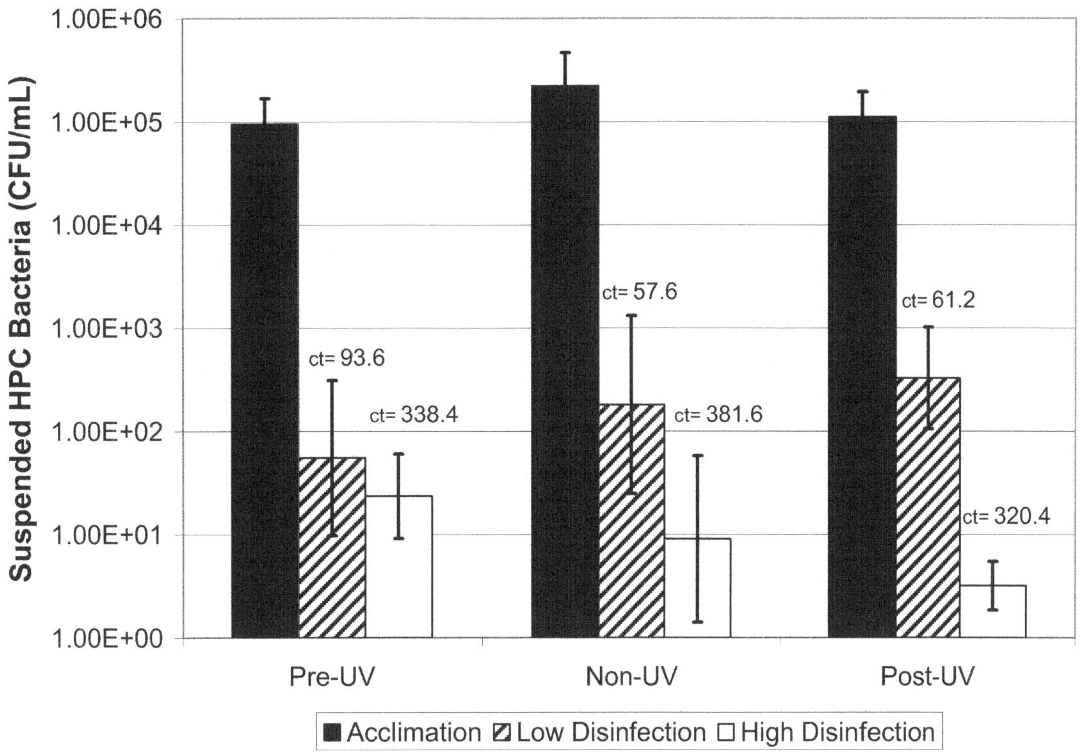

Figure 4.6 Suspended heterotrophic bacteria in polycarbonate ARs during chlorine dioxide disinfection

Monochloramine Trial

During low chloramine disinfection, low log removals were observed in the cast iron ARs as it was difficult to achieve the target residual of 1.0mg/L (Figure 4.7 and Table 4.5). Increased log removals were observed during the high disinfection phase. They were comparable for all treatments and ranged from 2.42 to 2.74 log. The Non-UV treated AR maintained a significantly higher Ct than both the UV treated reactors. The UV reactors experienced effluent HPC reductions of 2.42 and 2.68 log in the Post –UV and Pre-UV treated ARs, respectively (Table 4.5). Despite the lower monochloramine residuals maintained, the UV reactors had equivalent reductions to the Non-UV treated reactors. These results were found to be significant.

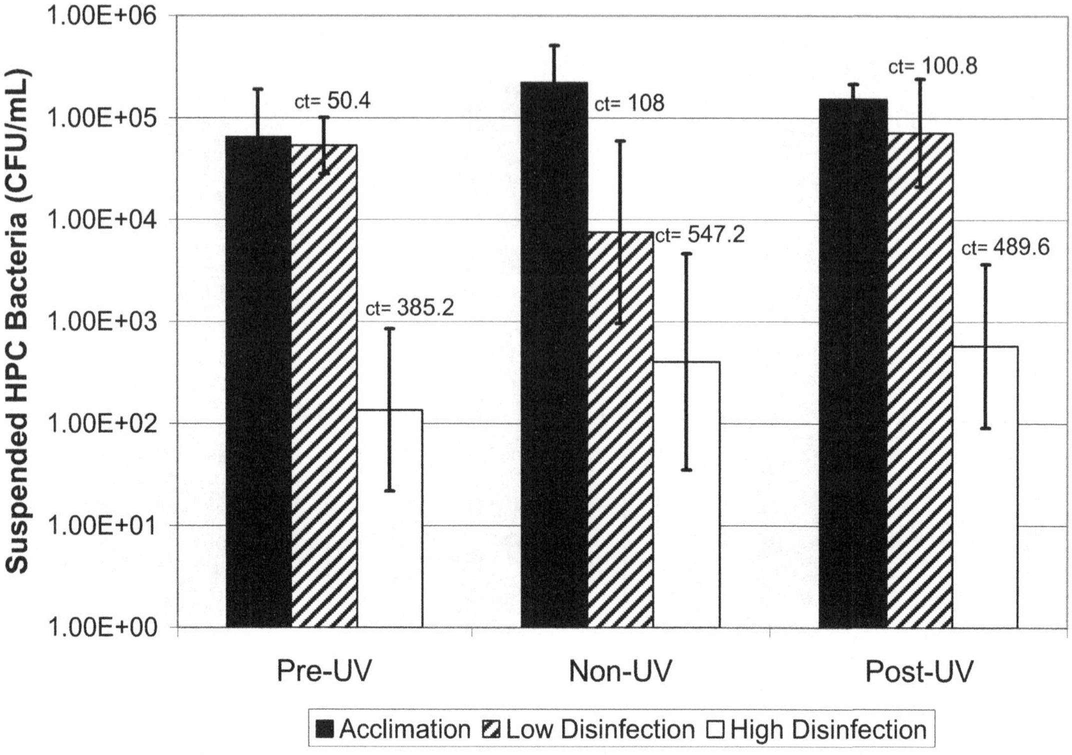

Figure 4.7 Suspended heterotrophic bacteria in cast iron ARs during chloramine disinfection

Maintaining and reaching the target monochloramine disinfectant residual was not as challenging in the ARs containing polycarbonate coupons as it was for those containing cast iron coupons. Effluent HPC reductions during low disinfection in the polycarbonate ARs spiked with *E. coli* K12 ranged from 1.23 to 1.89 log (Table 4.5). The Post-UV treated AR had the highest log removal of 1.89 and the lowest Ct of 216 (Figure 4.8). A similar trend was observed through high disinfection where the Pre-UV and Post–UV ARs exhibited larger effluent HPC reductions than the Non-UV treated system. Log reductions in the Pre and Post UV systems of 2.13 and 2.79 log corresponding with Ct values of 594 and 442.8, respectively, were notably greater than the 1.36 log reduction at a Ct of 594 for the Non-UV reactor (Figure 4.8). Nevertheless, there was no significant difference between effluent HPCs in Pre-UV, Non-UV and Post UV annular reactors treated with monochloramine.

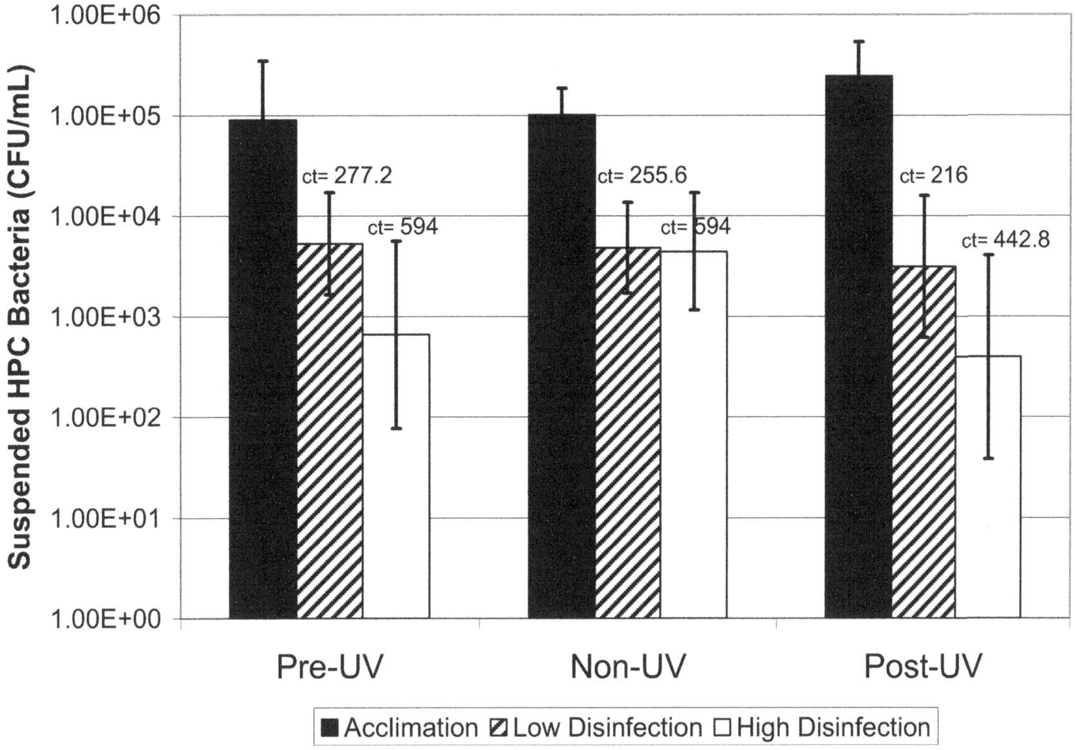

Figure 4.8 Suspended heterotrophic bacteria in polycarbonate ARs during chloramine disinfection

Biofilm Heterotrophic Bacteria

Biofilm heterotrophic bacteria log reductions for low and high disinfection are presented in Table 4.6 for Cl_2, ClO_2 and NH_2Cl, respectively. Log reductions were calculated using average HPC data from the acclimation, low disinfection and high disinfection phases.

Table 4.6
Biofilm HPC log reductions

Treatment	Biofilm HPCs- Log Reduction during Disinfection					
	Low Cl_2	High Cl_2	Low ClO_2	High ClO_2	Low NH_2Cl	High NH_2Cl
Pre-UV-CI - K12	0.18	-0.23	0.27	0.35	0.13	0.40
Pre-UV-CI-Mutant	-0.25	-0.55	0.52	0.14	0.07	0.58
Pre-UV-PC-K12	1.09	1.80	2.62	2.98	0.51	0.92
Pre-UV-PC-Mutant	1.51	1.71	1.30	2.61	0.27	1.08
CI - K12	-0.19	-0.58	0.20	0.17	0.79	1.04
CI-Mutant	-0.09	-0.37	0.28	0.20	0.11	0.51
PC-K12	1.20	2.53	2.15	2.26	1.23	1.22
PC-Mutant	1.76	2.61	-0.17	0.05	1.00	1.32
Post-UV-CI - K12	0.18	-0.17	0.51	0.14	0.37	1.45
Post-UV-CI-Mutant	-0.36	-0.16	-	-		
Post-UV-PC-K12	0.99	1.82	2.32	3.79	1.07	1.49
Post-UV-PC-Mutant	1.40	1.43	-	-		-

Effect of UV Disinfection

Figures 4.9- 4.14 illustrate the differences in effluent heterotrophic bacteria between Pre-UV, Non-UV and Post-UV for cast iron and polycarbonate ARs for each chemical disinfectant. In each figure, the corresponding Ct values during low and high disinfection are displayed.

Chlorine Trial. Overall, there was no statistically significant difference in biofilm HPC reduction between Pre-UV, Non-UV and Post-UV treated ARs spiked with *E. coli* K12 during chlorine disinfection (Figures 4.9 & 4.10). Interestingly, there was a significant difference in biofilm HPC reduction between the Pre-UV and Non-UV treated ARs that were spiked with the mutant *E. coli* strain. The p-values for ARs spiked with the mutant *E. coli* strain were $4.33 * 10^{-11}$ and $7.12 * 10^{-9}$ for cast iron and polycarbonate ARs, respectively.

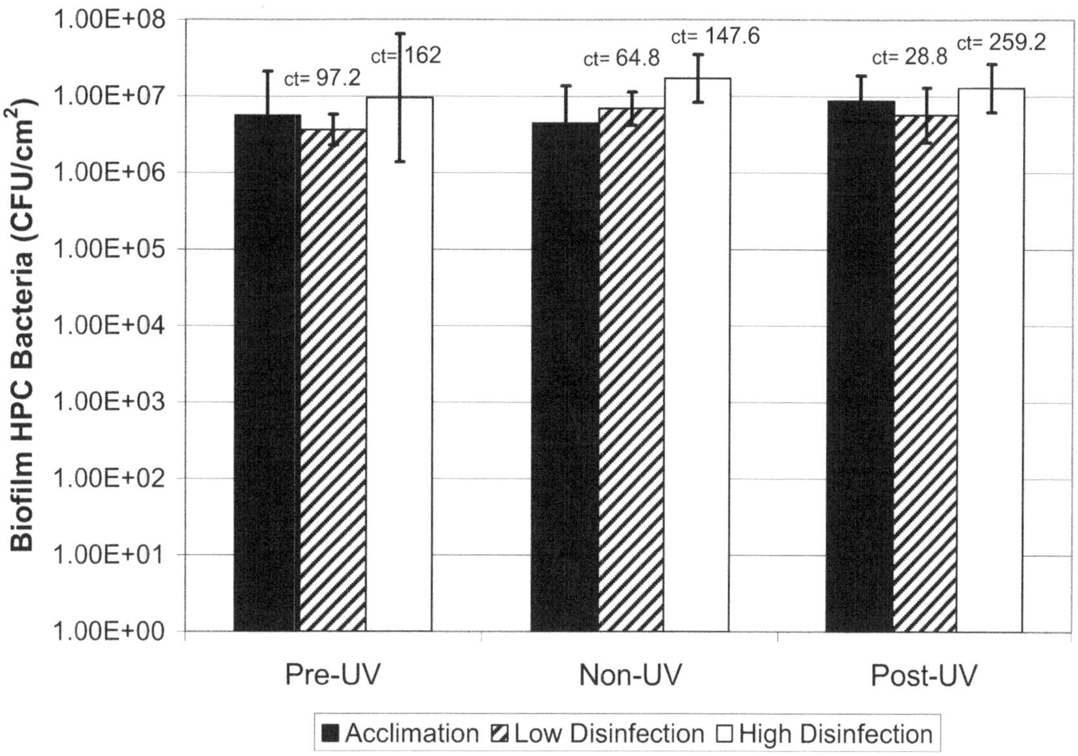

Figure 4.9 Biofilm heterotrophic bacteria in cast iron ARs during chlorine disinfection

In Figure 4.9, UV disinfection (Pre or Post) appeared to have no significant effect on biofilm HPCs in cast iron reactors. Chlorine also proved to be ineffective at reducing biofilm HPCs as HPC concentrations increased over time despite continuous disinfection. In polycarbonate reactors (Figure 4.10) during low disinfection the Non-UV treated AR had the greatest log removal of 1.20, followed by log removals of 1.09 and 0.99 in the Pre-UV and Post-UV systems, respectively. During high disinfection, the log removals were as follows: 2.53, 1.82 and 1.80 for Non-UV, Post-UV and Pre-UV reactors, correspondingly. The Non-UV AR maintained a higher Ct than the UV treated reactors, therefore had the Ct been similar in all ARs it is plausible that the UV treated ARs would have shown greater log reductions.

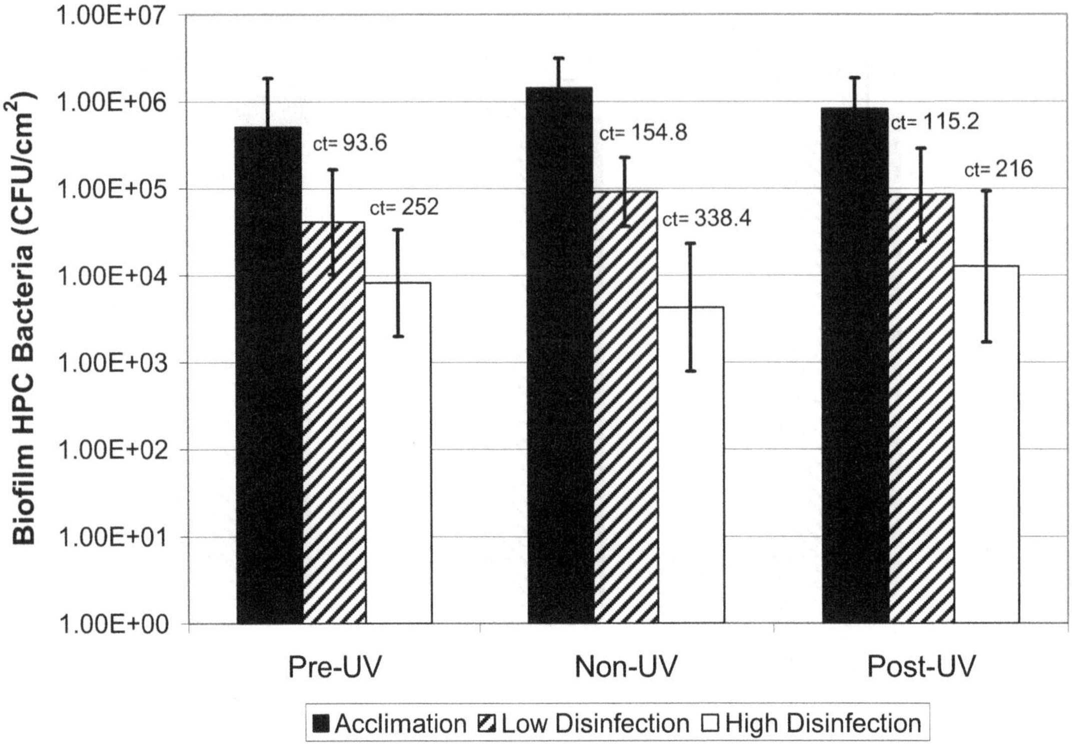

Figure 4.10 Biofilm heterotrophic bacteria in polycarbonate ARs during chlorine disinfection

Chlorine Dioxide Trial. Overall, in the cast iron reactors, there were no significant differences between biofilm HPCs in the UV treated ARs (Pre and Post) and the Non UV treated ARs. Log reductions during low disinfection were 0.51, 0.27 and 0.20 for Post-UV, Pre-UV and Non-UV, in that order. Interestingly, during high disinfection the Pre-UV treated AR had a higher biofilm HPC reduction (0.35 log) than the Non-UV treated AR (0.17 log) even though the Non-UV reactor maintained a much higher disinfectant residual (Ct=424.8) than the Pre-UV reactor (Ct= 284.4).

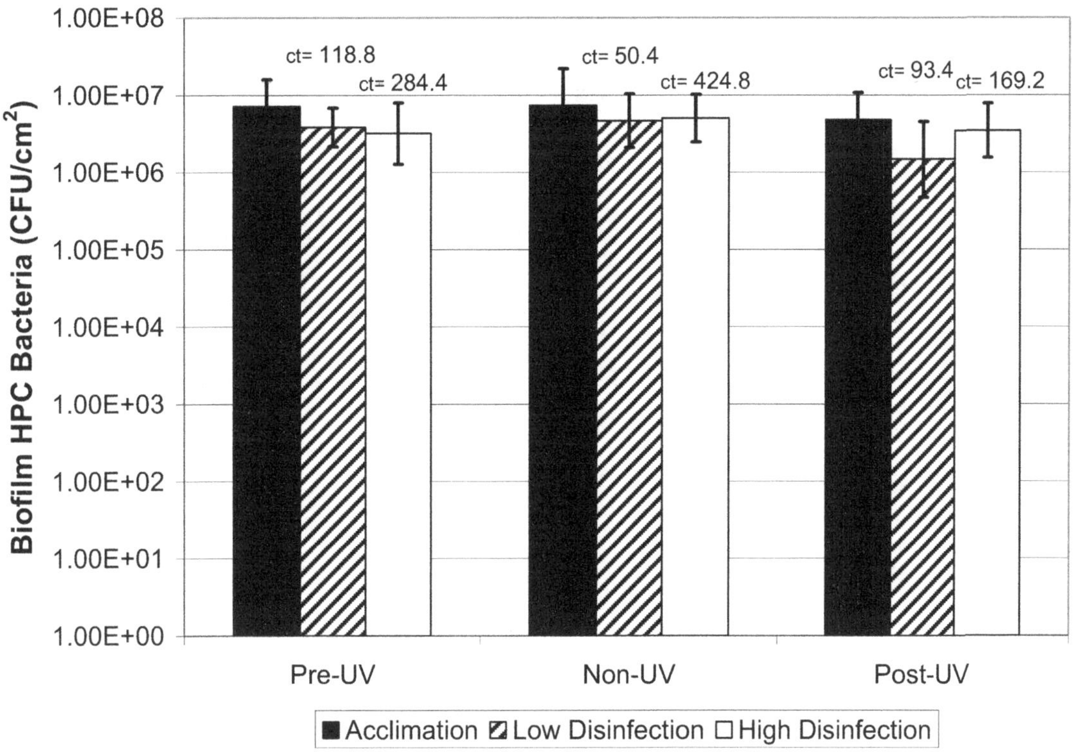

Figure 4.11 Biofilm heterotrophic bacteria in cast iron ARs during chlorine dioxide disinfection

In the reactors containing polycarbonate coupons, the UV treated reactors had the highest biofilm HPC log removal. During low disinfection the log removals in descending order were 2.62, 2.32 and 2.15 for Pre-UV, Post-UV and Non-UV ARs, respectively. The log reductions during high disinfection for the Post-, Pre- and Non-UV reactors were 3.79, 2.98 and 2.26 in that order. In high disinfection, the Non-UV treated reactor maintained the largest disinfectant residual (Ct= 381.6) however had the lowest log reduction. This same trend was observed for the cast iron ARs. In both cases, the highest Ct value did not always correspond to the largest log reduction. The ARs treated with UV (Pre and Post) realized greater log removals at lower chlorine dioxide concentrations than the reactor only treated with chemical disinfectant. Overall, there was no statistically significant difference between the Pre and Post- UV polycarbonate ARs in terms of biofilm HPC concentrations. However, the difference between the Pre-UV and Non-UV ARs had significance ($p > 0.001$).

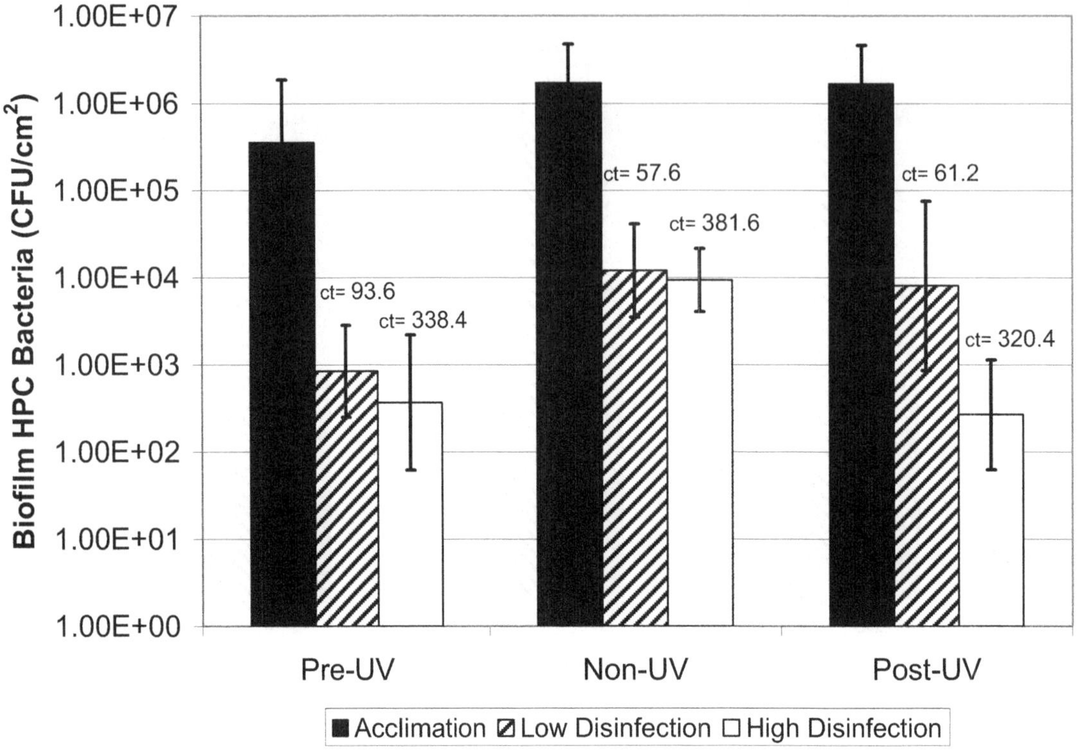

Figure 4.12 Biofilm heterotrophic bacteria in polycarbonate ARs during chlorine dioxide disinfection

Chloramines Trial. In general there were no statistically significant differences in biofilm HPCs between Pre-, Non- and Post-UV treated cast iron annular reactors (Figure 4.13). Low chloramine disinfection achieved log reductions of 0.79 for the Non-UV AR, 0.37 for the Post-UV AR and 0.13 for the Pre-UV AR. In this case, the highest log reduction (0.79 log) corresponded with the highest Ct value.This was not the case in high disinfection. During high disinfection the Non-UV treated AR achieved the largest Ct of 547.2, however only had a biofilm HPC reduction of 1.04 log. During high disinfection, the Post-UV reactor, achieved the largest log removal of 1.45 which corresponded to the second highest Ct of 489.6.

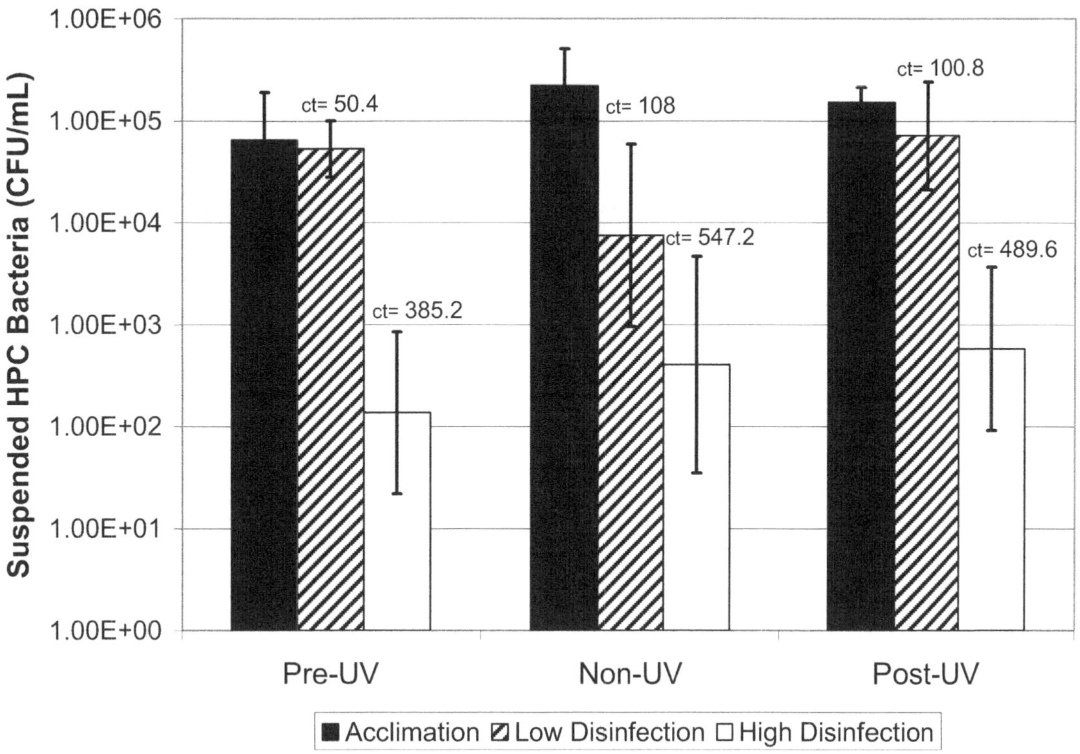

Figure 4.13 Biofilm heterotrophic bacteria in cast iron ARs during chloramine disinfection

In the systems containing polycarbonate coupons, the biofilm HPCs in the Pre-UV AR were slightly different than those in Non-UV AR (p=0.038) (Figure 4.14). Interestingly, the Pre-UV AR was slightly significantly different than the Post –UV AR (p= 0.02). Low chloramines disinfection yielded the following biofilm HPC log reductions: 1.23 (Non-UV), 1.07 (Post-UV) and 0.51 (Pre-UV). For high chloramines disinfection the trend found for cast iron reactors was also observed in the polycarbonate reactors; the largest log removal did not correspond with the highest Ct value (Figures 4.13 and 4.14). The log removals in high disinfection were 1.49, 1.22 and 0.92 for Post-, Non- and Pre-UV treated ARs. Although a Ct of 594 was maintained in both Pre- and Non-UV model distribution systems, a Ct of 442.8 was large enough in combination with Post-UV treatment to obtain a biofilm HPC reduction of 1.49 log.

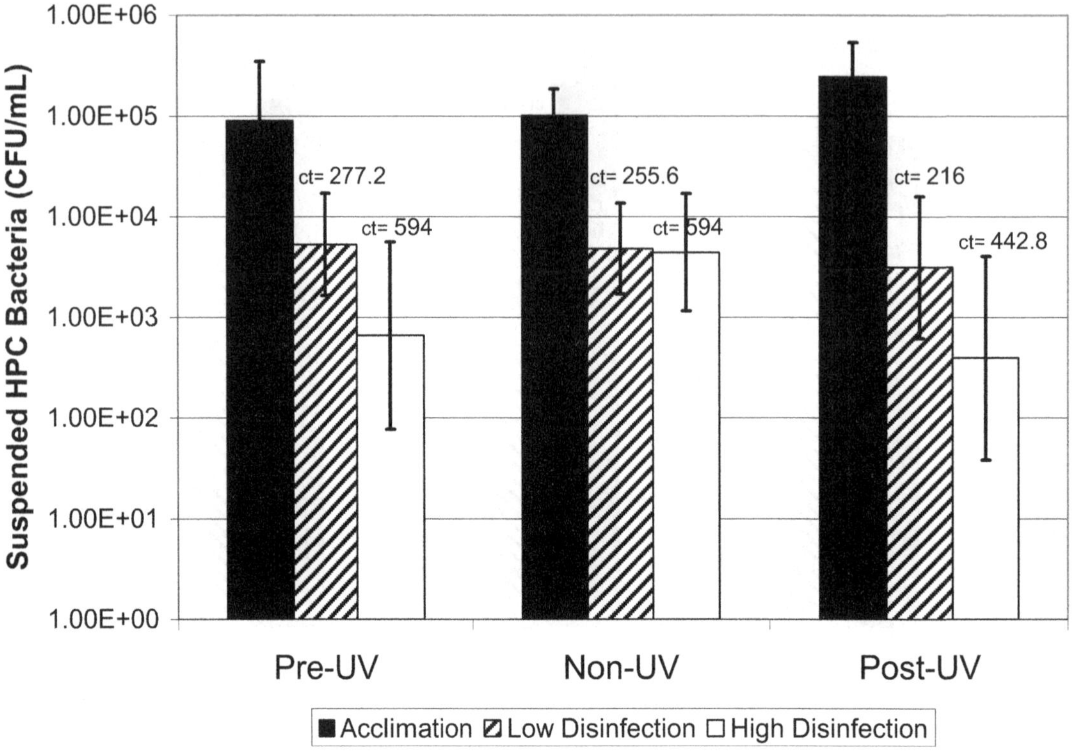

Figure 4.14 Biofilm heterotrophic bacteria in polycarbonate ARs during chloramine disinfection

Effluent *E. coli*

Effluent *E. coli* log reductions for low and high disinfection are presented in Table 4.7 for the Cl_2, ClO_2 and NH_2Cl trials, respectively. Log reductions were calculated using average *E. coli* data from the acclimation, low disinfection and high disinfection phases. Equation 4.3 was used to calculate log reduction.

$$\text{Log Reduction} = -\log \{\text{Avg. } E.\ coli_{disinfection} / \text{Avg. } E.\ coli_{acclimation}\} \quad (4.3)$$

After three to four weeks of acclimation, all reactors were spiked with either *E. coli* K12 or the MMR mutant strain as described in Chapter 3. The spiking was staggered over a few days to allow for *E. coli* to be spiked both directly into the Pre-UV and Non-UV reactors and then spiked through the UV lamp for the Post-UV ARs. After *E. coli* was added to the systems, a noticeable increase in effluent HPCs was observed (Table 4.7 and Figures 4.15-4.16). For both the chlorine and chlorine dioxide trials, the ARs were spiked a second time prior to disinfection because the *E. coli* had died off considerably.

Table 4.8 shows the changes in suspended heterotrophic bacteria after the first spike for all disinfectant trials. In most cases, effluent HPCs increased after the spike with exception of those ARs that were spiked with *E. coli* that received UV treatment (Post-UV) as well as AR4 and AR8 in the chlorine trial. It is probable that AR4 and AR8 had negative increases in HPCs because the spikes into these ARs were unsuccessful. *E. coli* counts were extremely low in these two reactors also explaining why they were re-spiked on day 59 (Figures 4.15-4.16).

Table 4.7
Effluent *E.coli* log Reductions

Treatment	*Effluent E. coli*- Log Reduction during Disinfection					
	Low Cl_2	High Cl_2	Low ClO_2	High ClO_2	Low NH_2Cl	High NH_2Cl
Pre-UV-CI - K12	2.62	2.62	2.47	3.38	2.53	2.67
Pre-UV-CI-Mutant	3.33	3.33	2.70	2.94	2.72	2.93
Pre-UV-PC-K12	3.38	3.38	3.00	3.38	2.95	2.90
Pre-UV-PC-Mutant	3.38	3.38	3.35	3.38	2.95	3.22
CI - K12	2.64	2.95	2.65	2.89	2.76	2.86
CI-Mutant	2.66	2.88	1.49	2.18	2.17	2.71
PC-K12	3.31	3.31	3.12	3.38	2.62	2.62
PC-Mutant	3.07	3.38	3.07	0.32	3.14	3.14
Post-UV-CI - K12	0.55	0.55	-0.19	0.19	0.58	0.58
Post-UV-CI-Mutant	-0.01	0.00	-	-	-	-
Post-UV-PC-K12	0.52	0.79	0.48	0.00	0.45	0.45
Post-UV-PC-Mutant	1.18	1.18	-	-	-	-

Table 4.8
Changes in effluent HPC concentrations after the *E. coli* spikes

Treatment	Effluent HPCs- Log Increase after *E. coli* spike		
	Cl_2	ClO_2	NH_2Cl
Pre-UV-CI - K12	0.28	0.33	1.50
Pre-UV-CI-Mutant	0.64	0.79	0.05
Pre-UV-PC-K12	0.49	0.46	1.42
Pre-UV-PC-Mutant	-0.38	0.92	0.08
CI - K12	1.88	0.73	0.48
CI-Mutant	2.30	0.58	1.29
PC-K12	1.21	0.76	0.04
PC-Mutant	-0.07	0.78	0.20
Post-UV-CI - K12	0.68	0.09	-0.42
Post-UV-CI-K12	0.70	-	-
Post-UV-PC-K12	0.60	0.68	0.08
Post-UV-PC-K12	-0.83	-	-

Overall, the increases in HPCs varied from one AR to the next and from one disinfectant trial to another with no evident trends. It is likely that the increases varied because it was impossible to exactly replicate each spike bottle. *E. coli* colonies were likely lost in the tubing as they were pumped into the reactors. In addition, it is possible that the die-off of colonies from one bottle to the next during the 24 hour spiking period varied.

In Figures 4.15 and 4.16, the evolution of effluent HPCs for polycarbonate and cast iron ARs during the chlorine trial are presented. These reactors were spiked with *E. coli* K12. The arrows on these figures indicate when the *E. coli* spikes occurred. Similar trends were observed for both the chlorine dioxide and chloramines disinfectant trials.

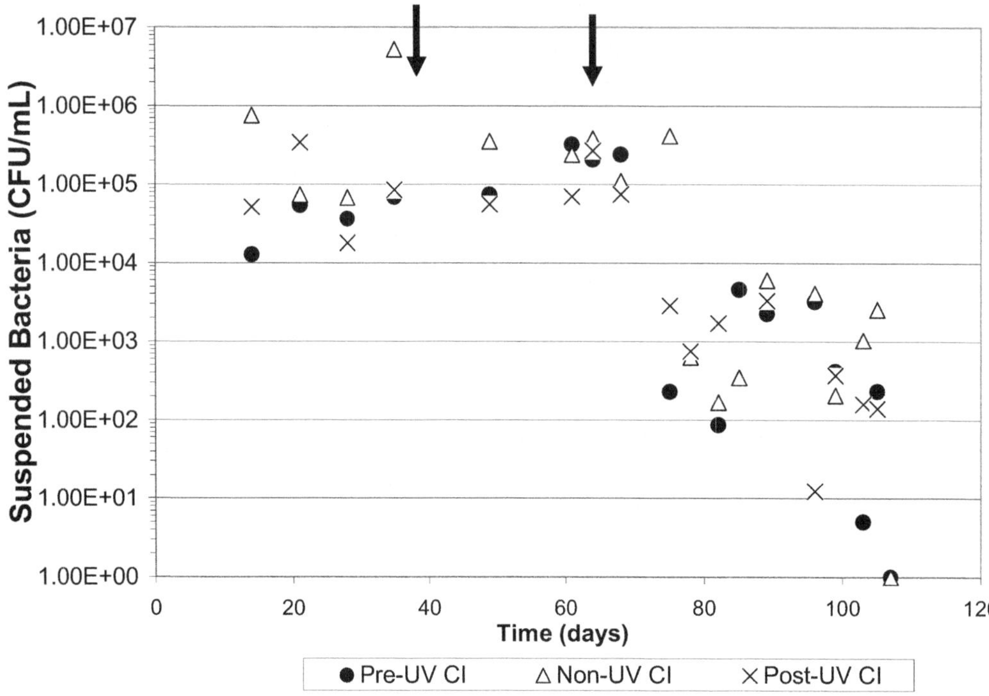

Figure 4.15 Effluent HPCs for cast iron ARs spiked with *E. coli K12* during the chlorine trial (arrows indicate when spikes occurred)

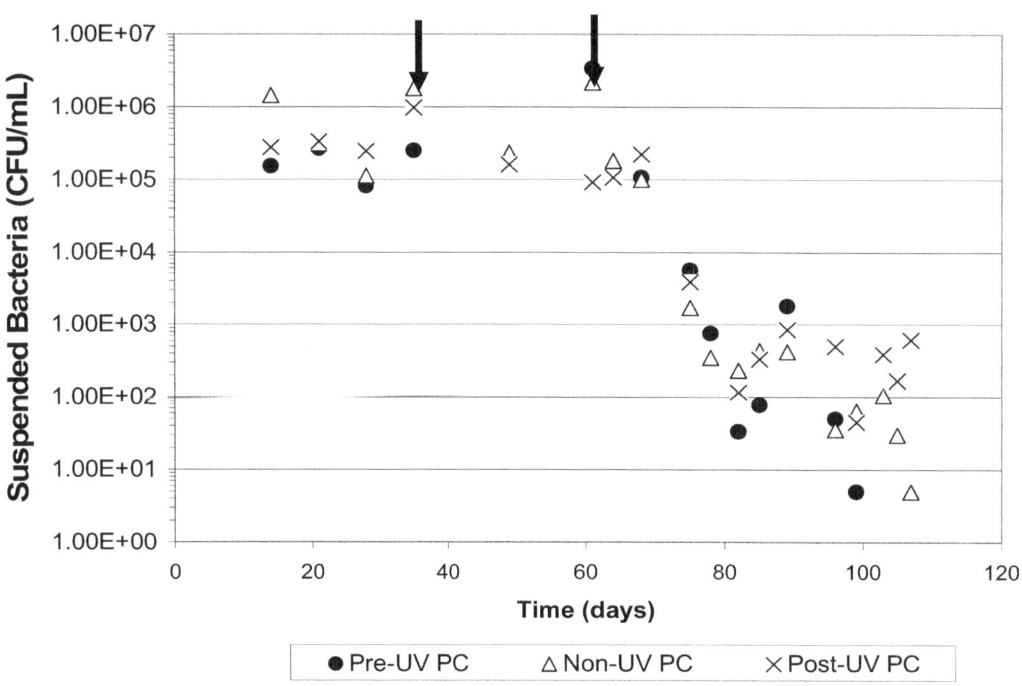

Figure 4.16 Effluent HPCs for polycarbonate ARs spiked with *E. coli K12* during the chlorine trial

Differences between E. coli K12 and Mutant strain

In terms of suspended *E. coli*, there were no significant differences between ARs that were spiked with *E. coli K12* and those that were spiked with the mutant (Table 4.9). However, the wild *E. coli* strain was more resistant to UV disinfection than its mutant counterpart. No mutant colonies were ever recovered Post-UV disinfection. Several spikes were conducted during the chlorine trial in an effort to see if any mutants would survive UV irradiation, however all trials were unsuccessful. As a result, after the chlorine trial, two reactors were removed from the setup (Post-UV-CI-Mutant and Post- UV-PC-Mutant).

Table 4.9
P-values for the difference between *E. coli K12* and Mutant strain on suspended *E. coli* counts in cast iron and polycarbonate ARs

Phase	Disinfectant	Cast Iron		Polycarbonate	
		Pre-UV	Non-UV	Pre-UV	Non-UV
Acclimation	Cl_2	0.248	0.882	1.00	0.374
	ClO_2	0.184	0.756	0.516	0.811
	NH_2Cl	0.595	0.293	0.444	0.431
Disinfection	Cl_2	1.00	0.736	1.00	0.331
	ClO_2	0.175	0.149	1.00	1.00
	NH_2Cl	0.899	0.516	0.430	1.00

Effect of UV disinfection on E. coli

As mentioned in Chapter 3 Material and Methods, three UV treatments were explored throughout the different disinfectant trials (Cl_2, ClO_2 & NH_2Cl). Pre-UV treatment means that UV disinfection occurred prior to the *E. coli* spikes and Post-UV treatment means the UV disinfection occurred after the *E. coli* spikes. In addition, four ARs in each run received no UV treatment (Non-UV). Figures 4.17-4.22 illustrate average suspended *E. coli* K12 counts in cast iron and polycarbonate ARs for all three experimental phases: acclimation, low disinfection and high disinfection, for each chemical disinfectant run.

Chlorine Trial. From Figures 4.17 and 4.18 it is evident that *E. coli* was significantly affected by UV disinfection (Post-UV). After the spikes there was less than 10 CFU/ 100mL of *E. coli* K12 in both the cast iron and polycarbonate Post-UV reactors. Therefore, Post-UV treatment did prove to be statistically superior against suspended *E. coli K12* than Non-UV or Pre-UV treatment during the acclimation and disinfection phases on both pipe materials.

There were no statistically significant differences between ARs treated with UV before the spike (Pre-UV) and those that were not UV treated. Surprisingly, in Figure 4.30 the *E. coli* in the Post UV reactor survived longer during low disinfection than those that were not UV disinfected. At Ct values of 93.6 and 154.8, log removals of 3.38 log and 3.37 log were attained for the Pre-UV and Non-UV reactors, respectively. For the Post-UV AR, a Ct of 115.2 was maintained, however only a 0.52 log reduction was reached. Perhaps, the *E. coli* colonies that survived UV disinfection repaired themselves and became more adaptable/ resistant and therefore were capable of surviving chlorine disinfection (Zimmer and Slawson 2002).

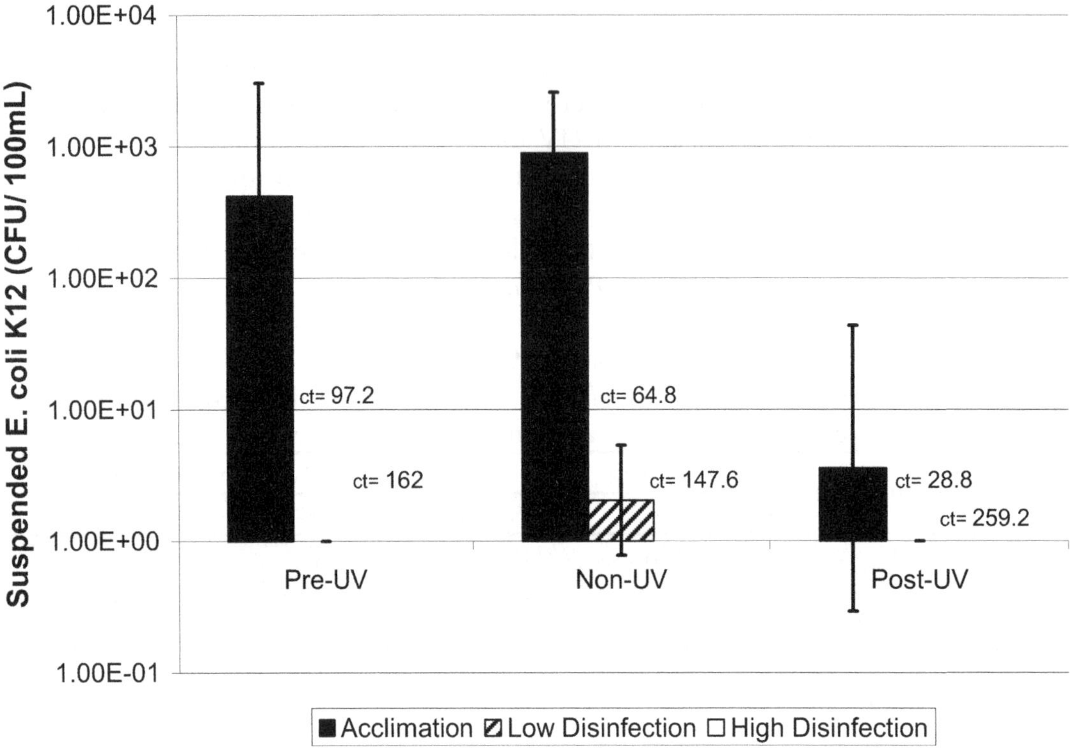

Figure 4.17 Suspended *E. coli K12* in cast iron ARs during chlorine disinfection

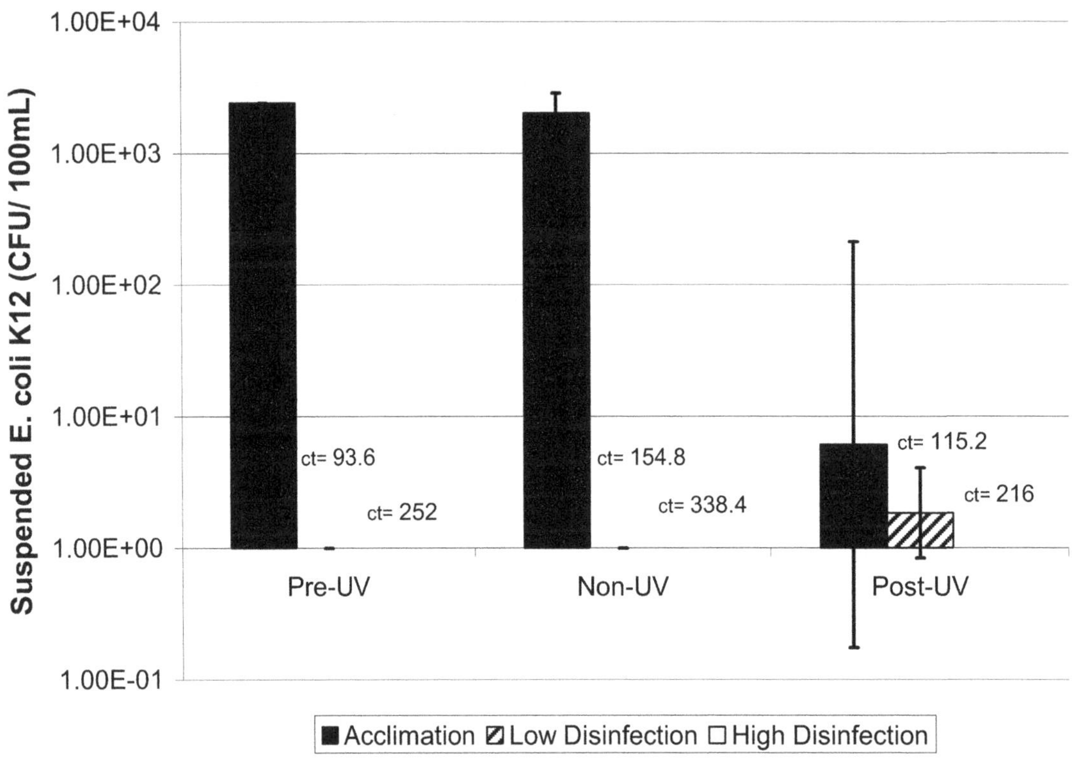

Figure 4.18 Suspended *E. coli K12* in polycarbonate ARs during chlorine disinfection

Chlorine Dioxide Trial. UV irradiation significantly affected effluent *E. coli K12* during the acclimation phase of chlorine dioxide run. After the spikes, there was approximately a 2.90 to 3.38 log difference in effluent *E. coli* between Post and Pre/Non UV disinfection in both cast iron and polycarbonate systems (Figure 4.19-4.20). There were no statistically significant differences between Pre and Non UV treated systems. These findings were consistent with the results from the chlorine trial.

Interestingly, after UV irradiation, coliforms were not detectable in the cast iron AR (Figure 4.19); during low chlorine dioxide disinfection, counts increased in the effluent by 0.19 log. The opposite occurred in the polycarbonate reactor where an average of 0.48 log of *E. coli* were present in suspension Post-UV treatment and then were undetectable throughout low disinfection.

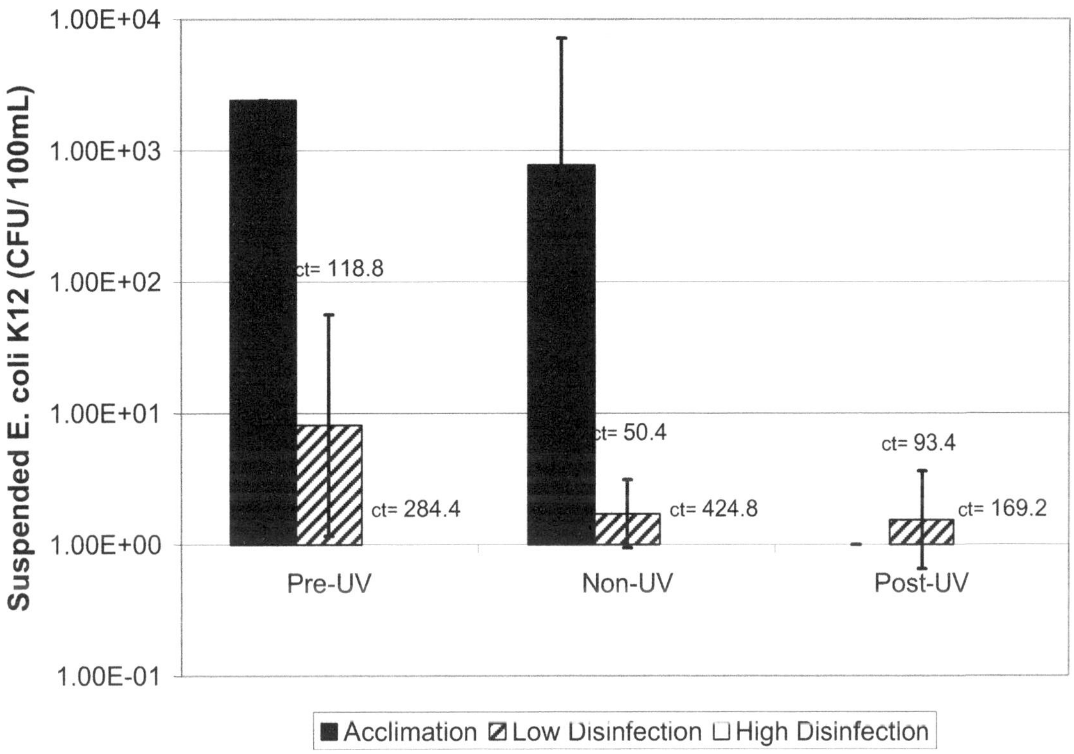

Figure 4.19 Suspended *E. coli K12* in cast iron ARs during chlorine dioxide disinfection

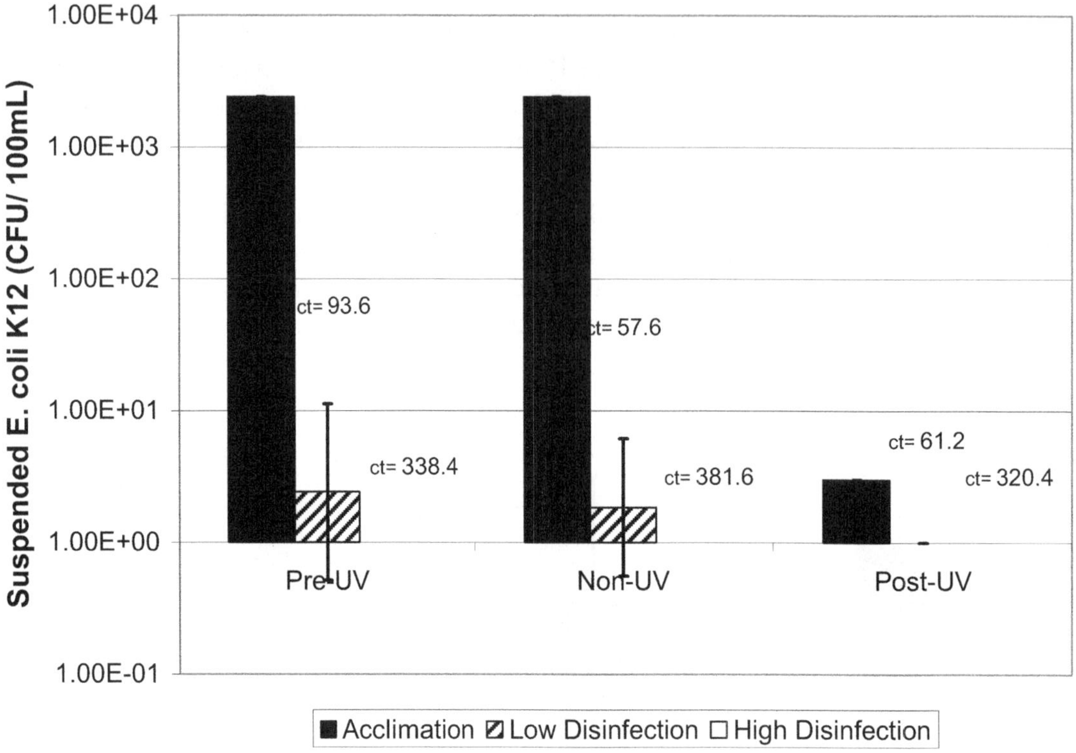

Figure 4.20 Suspended *E. coli K12* in polycarbonate ARs during chlorine dioxide disinfection

Chloramine Trial. Similar trends to those found in the chlorine and chlorine dioxide trials were also found for the monochloramine run (Figures 4.21-4.22). Again, there were no differences between *E. coli K12* counts for Pre-UV and Non-UV treated systems. Post-UV treatment decreased *E. coli* effluent counts by between 2.30 and 2.50 log when compared to the other two treatments. All of these results were found to be statistically significant.

Unlike both the chlorine and chlorine dioxide phases, low chloramines disinfection consistently decreased *E. coli* counts to below the detection limit in Post-UV reactors containing cast iron or polycarbonate coupons.

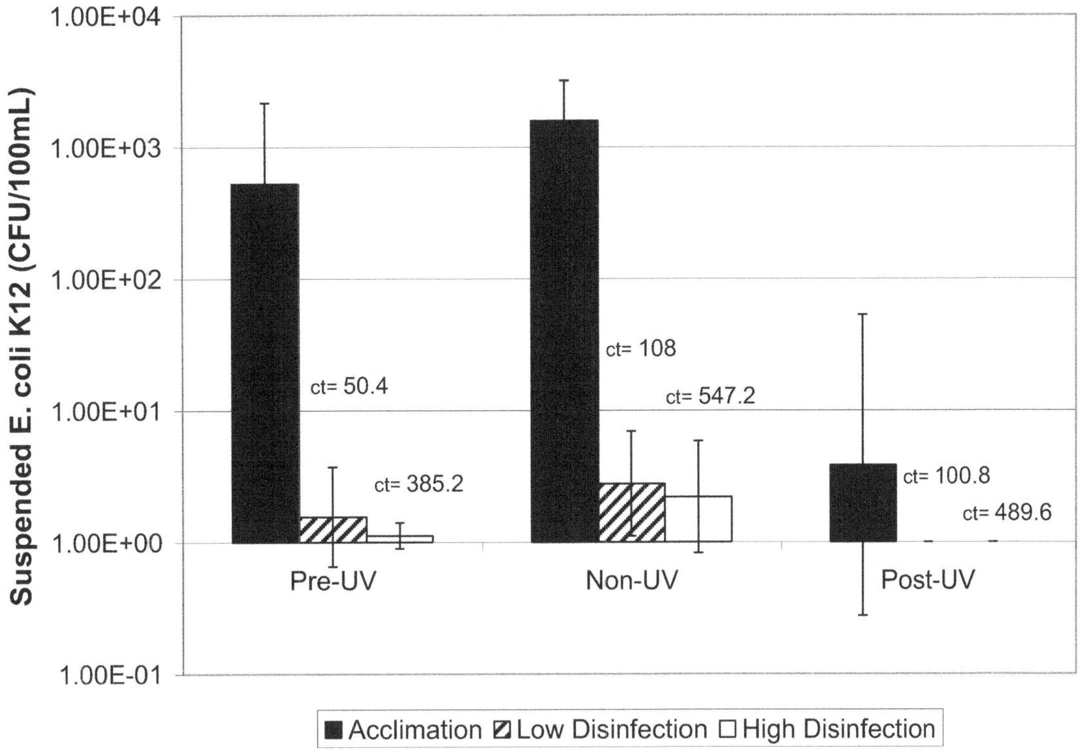

Figure 4.21 Suspended *E. coli K12* in cast iron ARs during chloramine disinfection

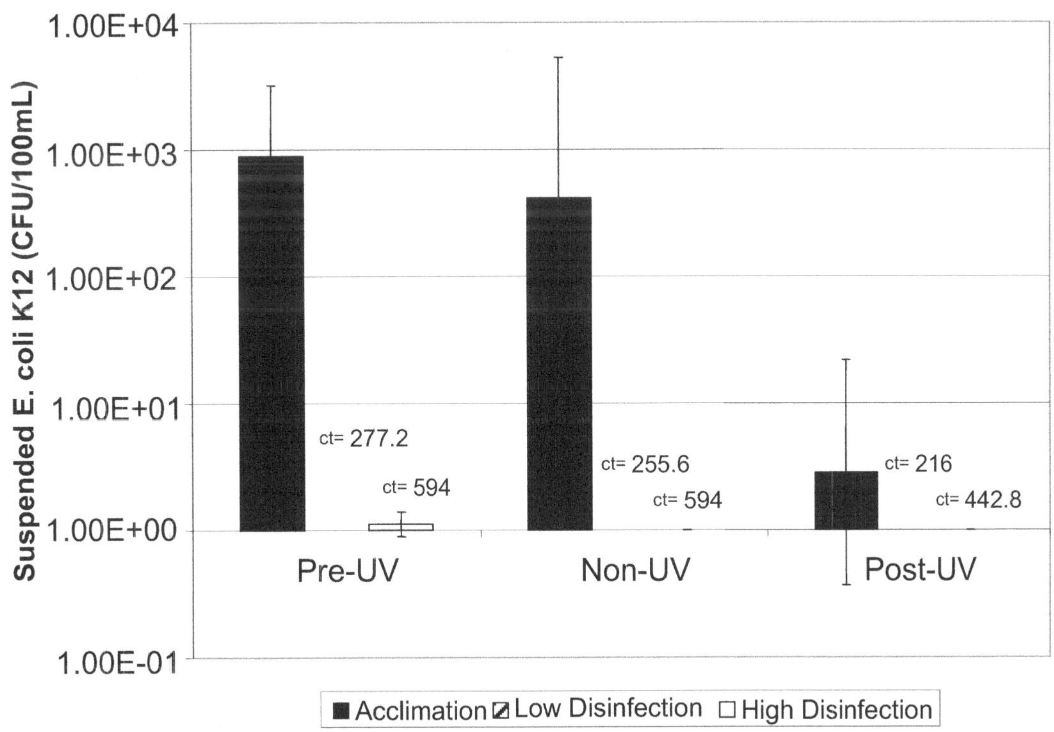

Figure 4.22 Suspended *E. coli K12* in polycarbonate ARs during chloramine disinfection

Analysis of Suspended *E. coli*

Overall, for each disinfectant trial there was no significant distinction between Pre and Non-UV disinfection in terms of suspended *E. coli* bacteria. However, the variation between *E. coli* that was Post –UV treated versus *E. coli* that was not UV irradiated was statistically significant for all disinfectant trials through the acclimation phases. Table 4.10 displays the ANOVA results when comparing the three UV groups. The shaded sections of the table indicate where the treatments were significantly different.

Table 4.10
P-values for the difference between UV treatments on suspended *E. coli* counts in cast iron and polycarbonate ARs

Phase	Disinfectant	Cast Iron		Polycarbonate	
		Pre- vs. Non-UV	Non- vs. Post-UV	Pre- vs. Non-UV	Non- vs. Post-UV
Acclimation	Cl_2	0.553	0.016	0.374	0.033
	ClO_2	0.272	0.005	1.00	0.005
	NH_2Cl	0.218	0.005	0.622	0.024
Disinfection	Cl_2	0.185	0.185	1.00	0.170
	ClO_2	0.182	0.898	0.791	0.327
	NH_2Cl	0.139	0.014	0.327	1.00

Effect of Pipe Material

In general, suspended *E. coli* in annular reactors containing cast iron did not vary considerably from those containing polycarbonate coupons (Table 4.11) with the exception of two cases. In the first case, coliforms were non-detect Post-UV disinfection in the cast iron AR during the ClO_2 trial, while they were 0.48 log in the polycarbonate reactor (Figures 4.19 - 4.20, Table 4.11). In the second instance, suspended *E. coli* counts were higher in the CI reactor than the PC reactor during chloramines disinfection (highlighted values in Table 4.11). A plausible explanation for this is the occurrence of nitrification within the cast iron AR. An increase in heterotrophic bacteria and a decrease in disinfectant residual were observed which potentially allowed the coliforms to persist and survive longer (Norton and LeChevallier, 1997). The shaded sections of the table indicate where the treatments were significantly different.

Table 4.11
P-values for the difference between cast iron and polycarbonate coupons on suspended *E. coli* counts

Phase	Disinfectant	Pre- UV	Non-UV	Post-UV
Acclimation	Cl_2	0.159	0.228	0.821
	ClO_2	0.272	0.712	0.051
	NH_2Cl	0.614	0.356	0.866
Disinfection	Cl_2	1	0.332	0.170
	ClO_2	0.113	0.179	1
	NH_2Cl	0.348	0.014	1

Effect of Secondary Disinfectants

As mentioned in Chapter 3, three secondary chemical disinfectants: chlorine, chlorine dioxide and chloramines were examined separately in three trials at two different doses (low and high). Figures 4.23 and 4.24 illustrate the average changes in suspended *E. coli K12* from acclimation, low disinfection, and high disinfection for Cl_2, ClO_2 and NH_2Cl in cast iron and polycarbonate systems.

Cast Iron. In the cast iron ARs, log reduction after low disinfection for all three chemical disinfectants for Pre-UV and Non-UV reactors ranged from 2.53- 2.76 log (Table 4.7, Figure 4.23). UV disinfection in the chlorine and monochloramine trials yielded similar initial average *E. coli* counts (0.55-0.58 log). Through low disinfection, both Cl_2 and NH_2Cl reduced effluent *E. coli* so that it was non-detectable.

The Post-UV-ClO_2 AR exhibited opposite trends to those found for Cl_2 and NH_2Cl. Initially, there were no detectable coliforms in the AR effluent Post-UV disinfection, however during low ClO_2 disinfection coliforms became detectable (Figure 4.23 -average increase of 0.19 log). Therefore, it is probable that coliforms were present in the effluent following UV disinfection but they were undetected by the substrate-based most probable number (MPN) detection method (see Chapter 3). The *E. coli* may have been injured by the UV irradiation and therefore were not detectable by conventional substrate- based methods (McFeters et al. 1986a). Between the UV disinfection and prior to chemical disinfection, it is plausible that the coliforms repaired themselves in the annular reactors and then became detectable at the beginning of low chlorine dioxide disinfection (Zimmer and Slawson 2002, and Williams et al. 2003). Although there were some slight inconsistencies between suspended *E. coli* in Cl_2, ClO_2 and NH_2Cl treated systems, none of them proved to be statistically significant in cast iron ARs during disinfection.

Polycarbonate. Similar to the cast iron ARs, polycarbonate ARs had consistent effluent *E. coli K12* log reductions between chemical disinfectants in the Pre-UV and Non-UV reactors during low disinfection. Log decreases ranged from 2.62 to 3.38 log (Table 4.7, Figure 4.24). In Figure 4.24, both Pre-UV Cl_2 and ClO_2 reactors maintained the same Ct of 93.6. At this equivalent Ct, chlorine slightly outperformed chlorine dioxide with *E. coli* decreases of 3.38 log and 3.30 log, respectively. Nevertheless, these results were insignificant. Additionally, comparable log drops among chemical disinfectants in the Post-UV systems (0.45-0.52 log) were realized during low disinfection (Table 4.7, Figure 4.24). Overall, it was determined that chlorine, chlorine dioxide and chloramines were equivalent on the basis of suspended *E. coli* reduction in polycarbonate annular reactors. These results were found to be significant.

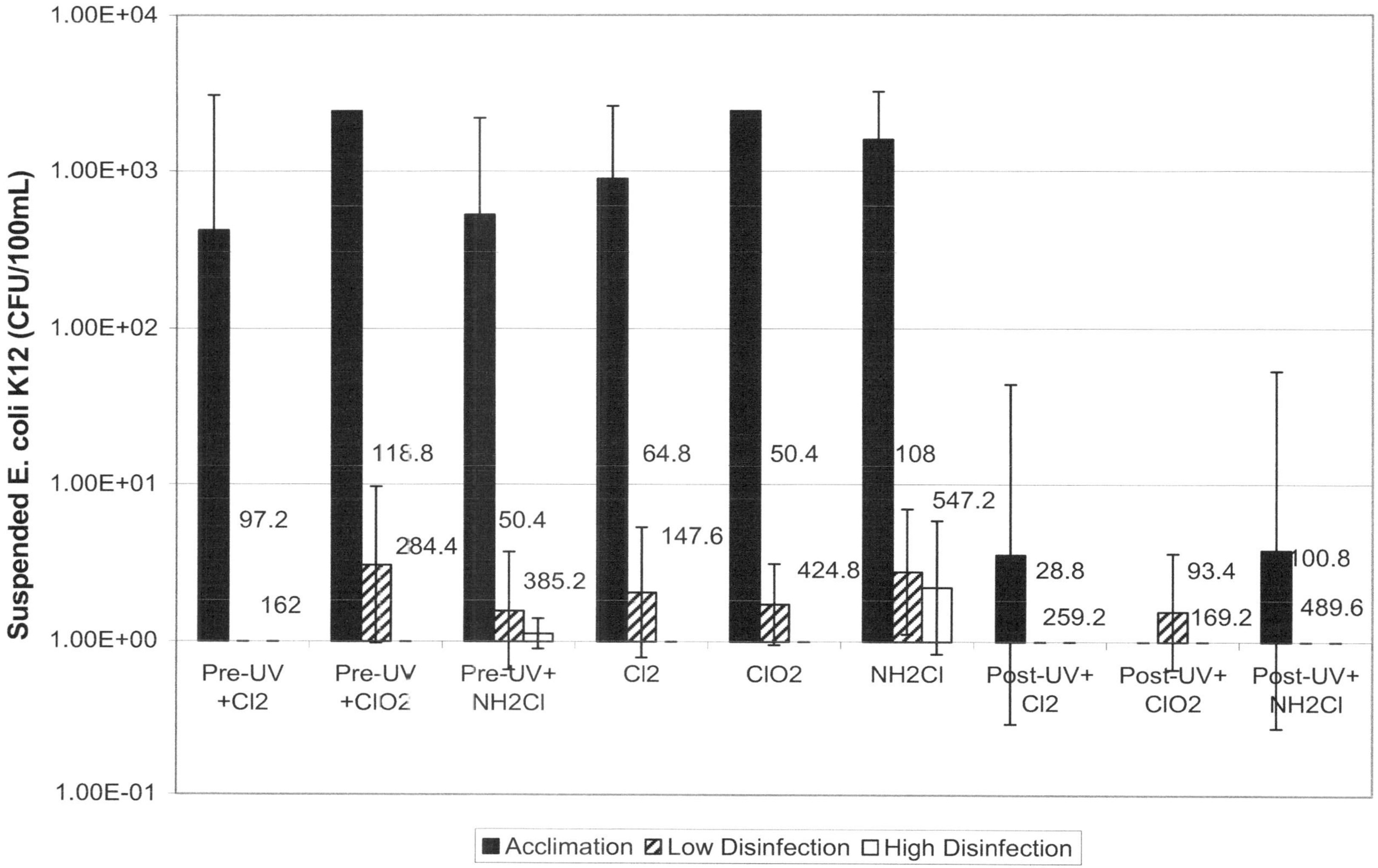

Figure 4.23 Suspended *E. coli K12* in cast iron ARs

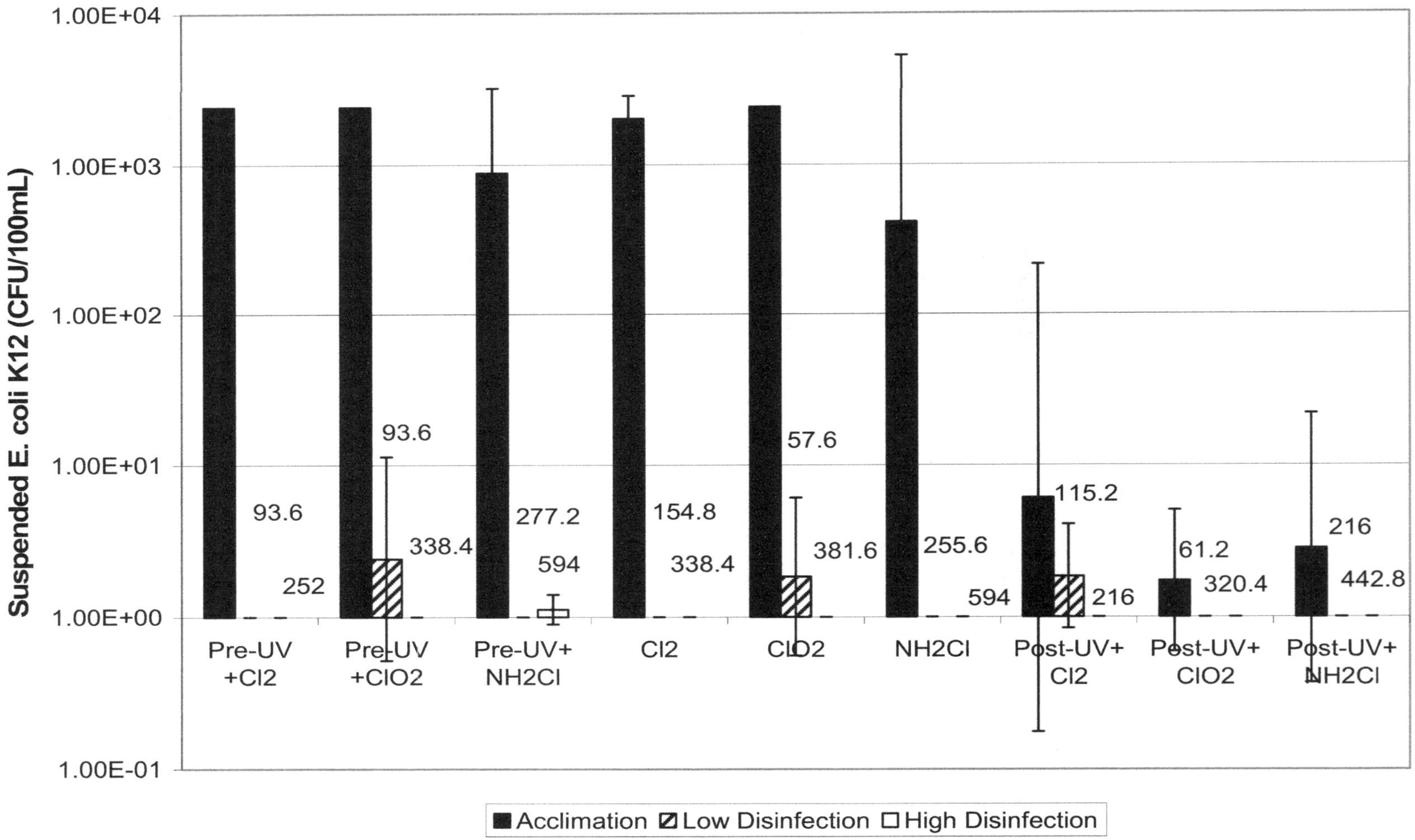

Figure 4.24 Suspended *E. coli K12* in polycarbonate ARs

Biofilm *E.coli*

Biofilm *E. coli* log reductions for low and high disinfection are presented in Table 4.12 for the Cl_2, ClO_2 and NH_2Cl trials, respectively. Log reductions were calculated using average *E. coli* data from the acclimation, low disinfection and high disinfection phases. Equation 4.3 was used to calculate log reduction.

Table 4.12
Biofilm *E. coli* log Reductions

Treatment	*Biofilm E. coli*- Log reduction during disinfection					
	Low Cl_2	High Cl_2	Low ClO_2	High ClO_2	Low NH_2Cl	High NH_2Cl
Pre-UV-CI - K12	3.53	4.03	2.52	5.03	3.46	3.79
Pre-UV-CI-Mutant	2.94	4.30	2.46	4.00	1.38	1.46
Pre-UV-PC-K12	3.89	4.47	4.05	4.36	1.85	2.11
Pre-UV-PC-Mutant	4.36	4.69	2.75	3.48	2.00	1.14
CI - K12	3.27	4.13	0.28	3.07	2.05	2.19
CI-Mutant	3.55	5.05	1.63	3.68	2.19	2.20
PC-K12	3.54	4.57	2.84	3.77	1.55	1.73
PC-Mutant	4.82	4.90	2.42	3.30	3.41	3.42
Post-UV-CI - K12	0.55	0.55	1.39	1.40	0.57	0.34
Post-UV-CI-Mutant	-0.01	0.00	-	-	-	-
Post-UV-PC-K12	0.09	0.30	0.00	0.00	0.54	0.45
Post-UV-PC-Mutant	-0.02	0.00	-	-	-	-

HPC increase due to E. coli Spike

As previously mentioned, all reactors were spiked with either *E. coli* K12 or the MMR mutant strain after three to four weeks acclimation (see Chapter 3). The spiking was staggered over a few days to allow for *E. coli* to be spiked both directly into the Pre-UV and Non-UV reactors and then spiked through the UV lamp for the Post-UV ARs. After *E. coli* was added to the systems, a noticeable increase in biofilm HPCs was observed (Table 4.13 and Figures 4.25-4.26). For both the chlorine and chlorine dioxide trials, the ARs were spiked a second time prior to disinfection because the *E. coli* had died off considerably.

Table 4.13 shows the changes in biofilm heterotrophic bacteria after the first spike for all disinfectant trials. In most cases, biofilm HPCs increased after the spike with exception of those ARs that were spiked with *E. coli* that received UV treatment (Post-UV), and AR7 during the Cl_2 and ClO_2 trials and AR8 in the Cl_2 run.

Overall, the increases in HPCs varied from one AR to the next and from one disinfectant trial to another with no evident trends. It is likely that the increases varied for the same reasons mentioned previously. In addition, the biofilm community from one AR to the next likely varied because several factors: influent water quality, pipe material, *E. coli* strain and/or the presence or absence of UV disinfection. Consequently, various dominant organisms may have developed in each reactor and possibly interfered with or encouraged the attachment of *E. coli*.

In Figures 4.25 and 4.26, the evolution of biofilm HPCs for cast iron and polycarbonate ARs during the chlorine trial are presented. These reactors were spiked with *E. coli* K12. The arrows on these figures indicate when the *E. coli* spikes occurred. Similar trends were observed for both the chlorine dioxide and chloramines disinfectant trials.

Table 4.13
Changes in biofilm HPC concentrations after the *E. coli* spikes

AR	Treatment	Biofilm HPCs- Log Increase after *E. coli* spike		
		Cl_2	ClO_2	NH_2Cl
1	Pre-UV-CI - K12	0.34	1.04	0.81
2	Pre-UV-CI-Mutant	0.13	0.68	0.77
3	Pre-UV-PC-K12	0.93	0.43	1.36
4	Pre-UV-PC-Mutant	0.57	1.98	0.33
5	CI - K12	0.98	0.27	0.14
6	CI-Mutant	0.39	0.68	0.91
7	PC-K12	-0.31	-0.60	0.55
8	PC-Mutant	-0.04	0.97	0.16
9	Post-UV-CI - K12	0.06	-0.33	-0.10
10	Post-UV-CI-Mutant	0.22		
11	Post-UV-PC-K12	0.26	0.85	-0.10
12	Post-UV-PC-K12	-0.09		

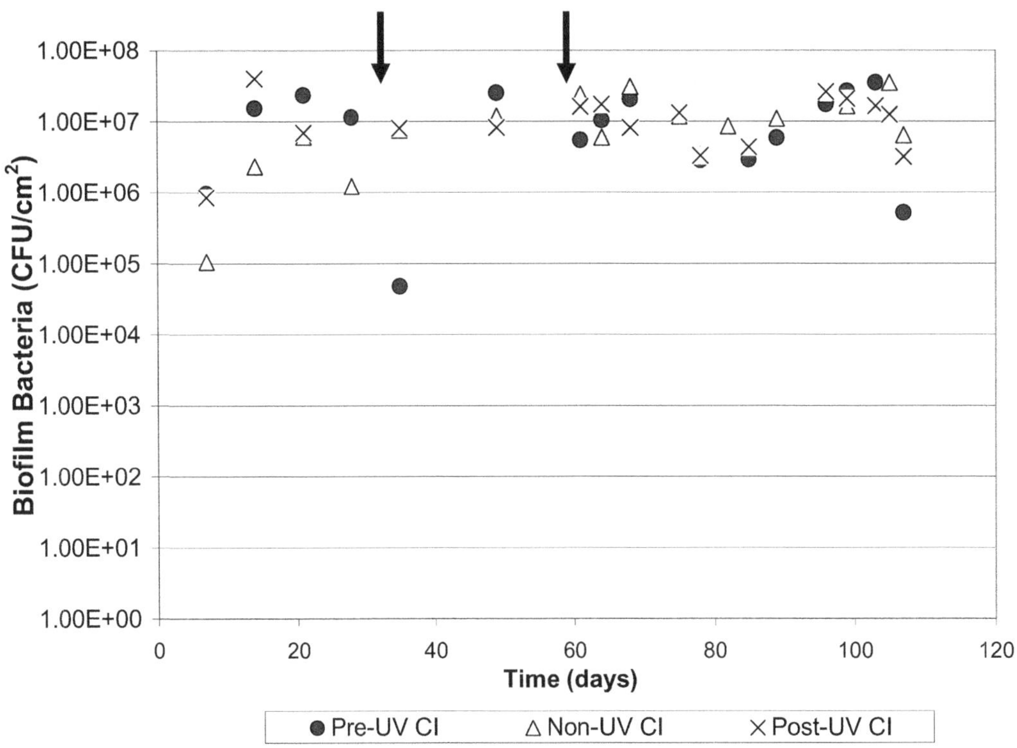

Figure 4.25 Biofilm HPCs for cast iron ARs spiked with *E. coli K12* during the chlorine trial (arrows indicate when spikes occurred)

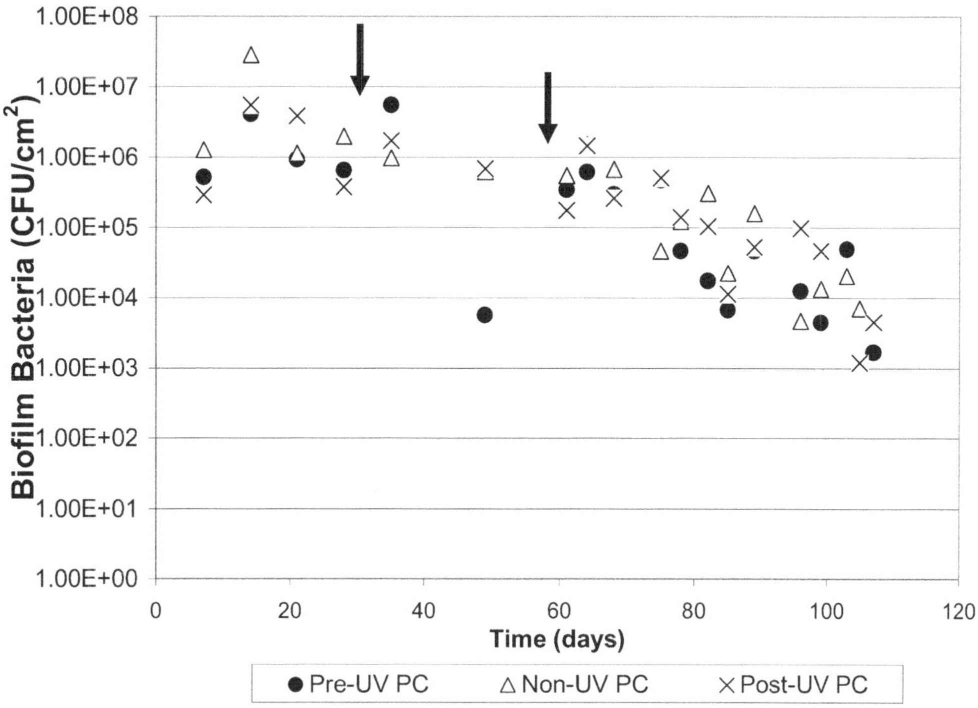

Figure 4.26 Biofilm HPCs for polycarbonate ARs spiked with *E. coli K12* during the chlorine trial (arrows indicate when spikes occurred)

Differences between E. coli K12 and Mutant strain

In terms of biofilm *E. coli* in Pre and Non-UV ARs, there were no significant differences between those that were spiked with *E. coli K12* and those that were spiked with the mutant during acclimation (Table 4.14). However in Post-UV ARs, the wild *E. coli* strain was more resistant to UV disinfection than its mutant counterpart. No mutant colonies were ever recovered from the biofilm Post-UV disinfection. Several spikes were conducted during the chlorine trial in an effort to see if any mutants would survive UV irradiation, however all trials were unsuccessful. As a result, after the chlorine trial, two reactors were removed from the setup.

In contrast, during chemical disinfection, on several occasions there were statistically significant differences in biofilm *E. coli* counts between the two *E. coli* strains (Table 4.14). Firstly, in the Pre-UV-CI- Cl2 ARs, the mutant survived longer than the K12. However, the chlorine residuals were also consistently higher in the AR containing *E. coli K12* than the AR with the *E. coli* mutant. Thus indicating that this variation was likely attributed to the inconsistent residuals between the two ARs. For the K12 and mutant ARs the Ct values for low disinfection were 97.2 and 54; for high disinfection they were 162 and 115.2, respectively.

The opposite was found for the Non-UV-CI-NH_2Cl ARs; the K12 strain yielded higher biofilm *E. coli* counts than its mutant counterpart. In this case, the disinfectant residual was equivalent for both strains during low disinfection and was higher in the K12 AR during high disinfection (Ct = 547.2 (K12) and 435.6 (mutant)). Therefore it is possible that *E. coli K12* was more resistant to NH_2Cl disinfection on cast iron surfaces than the *E. coli* mutant.

Similar results were found for biofilm *E. coli* counts on polycarbonate surfaces. In the Non-UV-PC-Cl_2 ARs, *E. coli K12* was more resistant to chlorine disinfection. In this instance, the residuals were similar between the ARs, therefore implying that the latter statement may be true (Table 4.2). Finally, in the Pre-UV-PC-NH_2Cl ARs, the AR spiked with the mutant strain yielded higher average counts during disinfection, yet it also maintained a lower disinfectant residual than the AR spiked with *E. coli K12*. The Ct values during high disinfection were 594 and 529.2, for the K12 and mutant reactors, in that order.

In conclusion, it appears as though *E. coli K12* was more resistant to chlorine and monochloramine disinfection on cast iron and polycarbonate surfaces than the *E. coli* mutant. These results are consistent with the effects of UV disinfection on the two *E. coli* species. In chlorine dioxide disinfection, the two types of *E. coli* behaved similarly and there were no significant differences. Nevertheless, more studies should be conducted to confirm the above findings.

Table 4.14
P-values for the difference between *E. coli K12* and the mutant strain on biofilm *E. coli* counts in cast iron and polycarbonate ARs (shaded areas indicate significance)

Phase	Disinfectant	Cast Iron		Polycarbonate	
		Pre-UV	Non-UV	Pre-UV	Non-UV
Acclimation	Cl_2	0.621	0.343	0.795	0.684
	ClO_2	0.402	0.400	0.965	0.489
	NH_2Cl	0.075	0.335	0.377	0.157
Disinfection	Cl_2	0.005	0.667	0.446	0.039
	ClO_2	0.054	0.080	0.282	0.425
	NH_2Cl	0.611	0.021	0.027	0.210

Effect of UV treatment on E. coli

Figures 4.27-4.32 illustrate average biofilm *E. coli* K12 counts in cast iron and polycarbonate ARs for all three experimental phases: acclimation, low disinfection and high disinfection, for each chemical disinfectant run.

Chlorine Trial. From Figures 4.27 and 4.28 it is evident that biofilm *E. coli* was significantly affected by UV disinfection (Post-UV). After the *E. coli* passed through the UV lamp, less than 10 CFU/ cm^2 attached themselves to cast iron surfaces, whereas, they were undetected on polycarbonate surfaces. In contrast, *E. coli* was established at quantities in the 10^4 CFU/cm^2 range and were statistically equal in the Pre-UV and Non-UV systems. Consequently, Post-UV treatment was statistically superior to Non-UV and Pre-UV treatment at reducing biofilm *E. coli* during the acclimation phase on both pipe materials.

In the disinfection stage, the Pre-UV-CI AR reduced on average more biofilm *E. coli* than the Non-UV-CI AR (Figure 4.27). These results were significant; however they are likely attributed to the higher chlorine residuals maintained in the Pre-UV AR and not the effect of Pre-UV treatment alone (Table 4.15). In the polycarbonate systems, there were no substantial variations between Pre-UV and Non-UV treatment throughout disinfection (Figure 4.28).

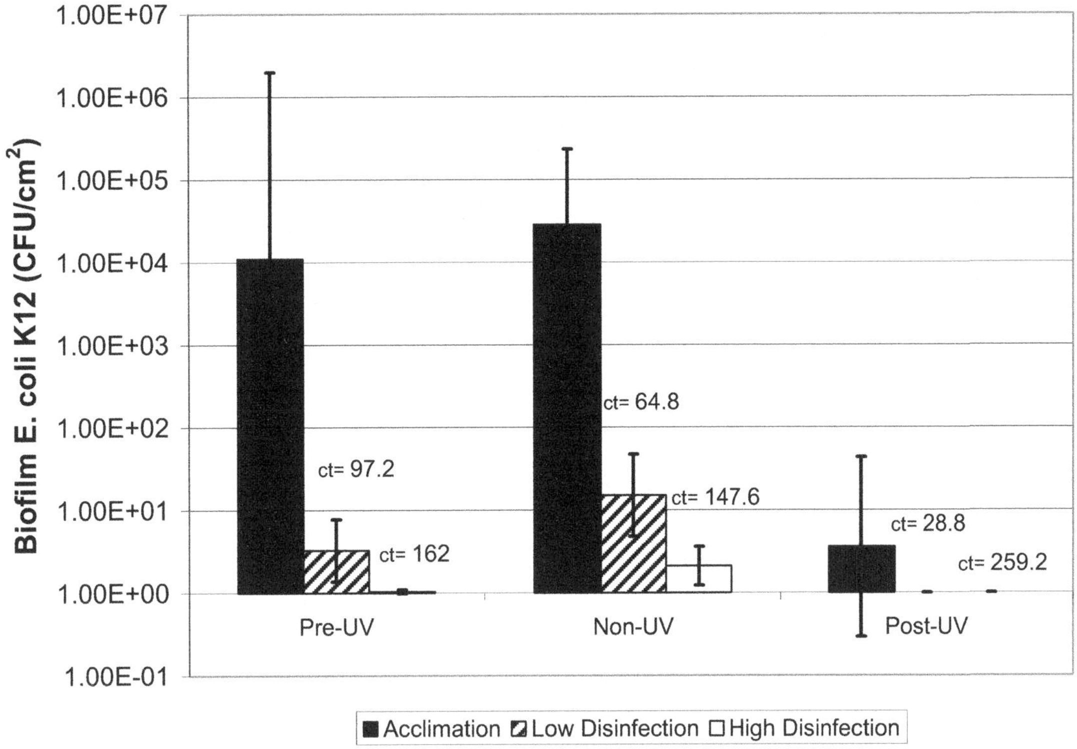

Figure 4.27 Biofilm *E. coli K12* in cast iron ARs during chlorine disinfection

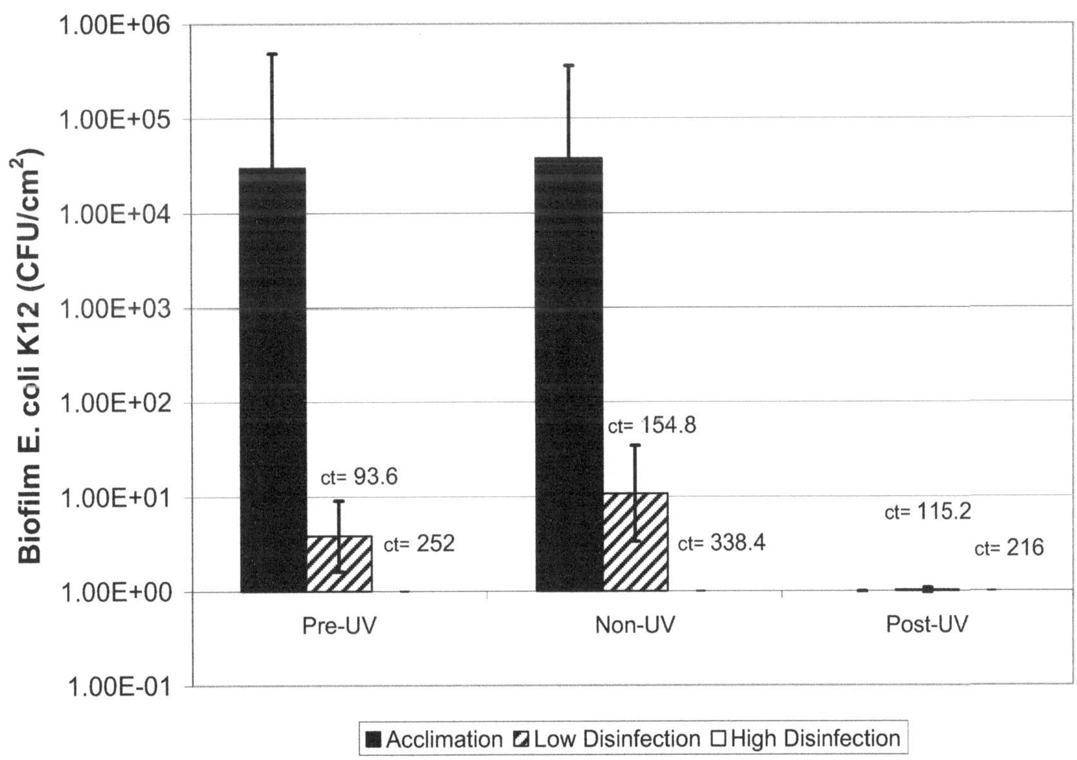

Figure 4.28 Biofilm *E. coli K12* in polycarbonate ARs during chlorine disinfection

Table 4.15
P-values for the difference between UV treatments on biofilm *E. coli* counts in cast iron and polycarbonate ARs

Phase	Disinfectant	Cast Iron		Polycarbonate	
		Pre- vs. Non-UV	Non- vs. Post-UV	Pre- vs. Non-UV	Non- vs. Post-UV
Acclimation	Cl_2	0.758	0.006	0.906	0.002
	ClO_2	0.294	0.302	0.515	0.005
	NH_2Cl	0.245	0.057	0.438	0.113
Disinfection	Cl_2	0.049	0.001	0.370	0.089
	ClO_2	0.530	1.44×10^{-5}	0.273	0.083
	NH_2Cl	0.054	0.037	0.331	0.362

Chlorine Dioxide Trial. UV irradiation significantly affected effluent *E. coli K12* during the acclimation phase of chlorine dioxide run. Similar to the chlorine trial, after the spikes there was approximately a 2.00 to 4.40 log difference in biofilm *E. coli* between Post and Pre/Non-UV disinfection in both cast iron and polycarbonate systems (Figure 4.29-4.30). However these results were only significant for the polycarbonate ARs and not for the cast iron ARs (Table 4.15). There were no statistically significant differences between Pre and Non UV treated systems during acclimation. These findings were consistent with the results for effluent *E. coli* in the chlorine dioxide trial. As disinfection proceeded there were substantial variations between the Non-UV –CI AR and the Post-UV-CI AR (p= 1.44×10^{-5}).

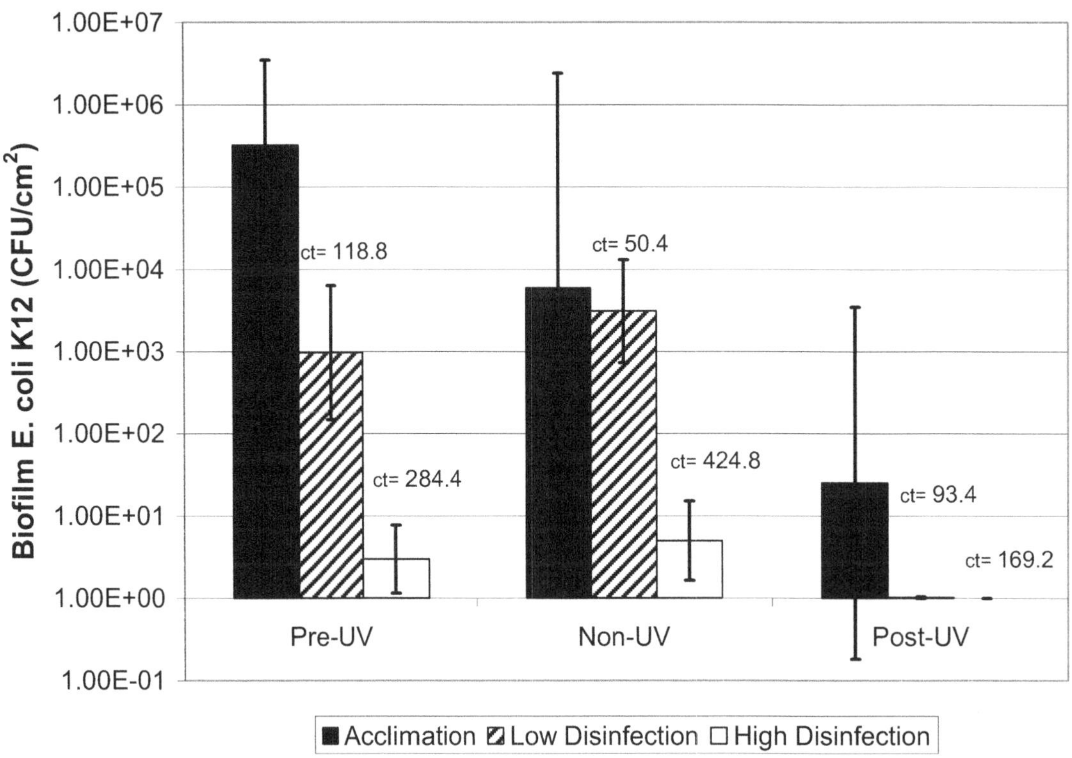

Figure 4.29 Biofilm *E. coli K12* in cast iron ARs during chlorine dioxide disinfection

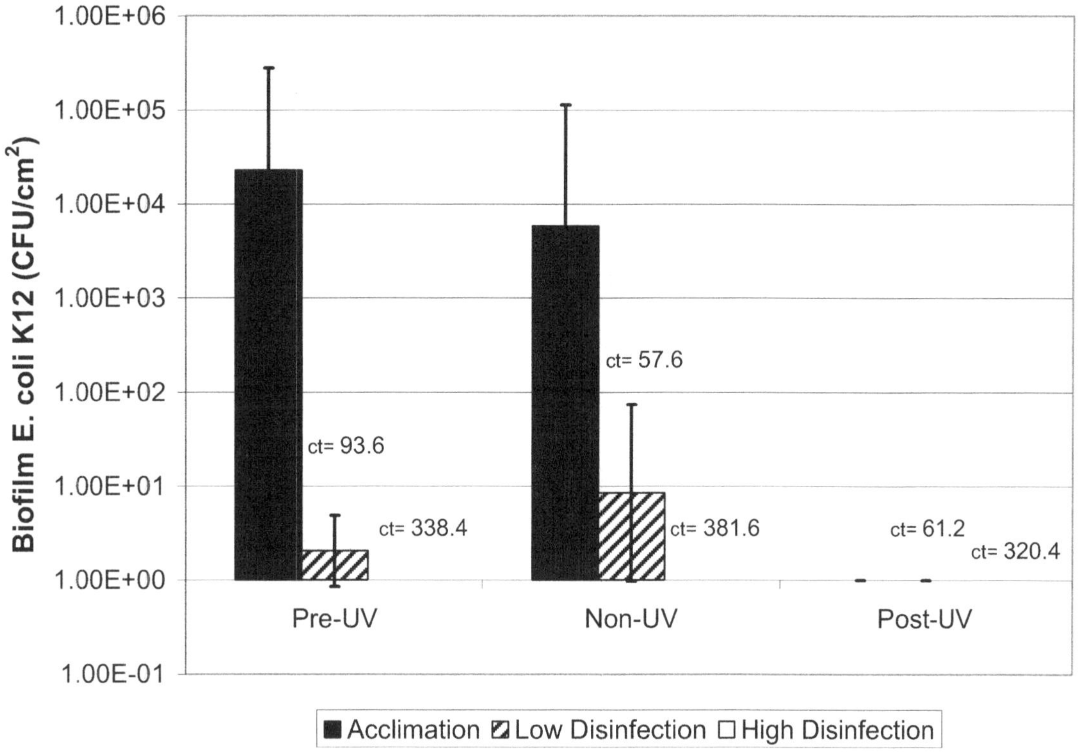

Figure 4.30 Biofilm *E. coli K12* in polycarbonate ARs during chlorine dioxide disinfection

Chloramine Trial. Similar trends to those found in the chlorine and chlorine dioxide runs were also found for the monochloramine trial (Figures 4.31-4.32). Again, there were no differences between *E. coli K12* counts for Pre-UV and Non-UV treated systems during acclimation. Post-UV treatment decreased *E. coli* effluent counts by between 1.5 and 3.20 log when compared to the other two treatments. Strangely, these results were not found to be statistically significant. A likely explanation for this would be that the initial *E. coli* establishment within the biofilm of the Non-UV and Pre-UV reactors was not as high as in the chlorine and chlorine dioxide trials.

Biofilm *E. coli K12* counts were significantly larger in the cast iron Non-UV reactor than the Pre-UV system throughout disinfection. Despite the fact that the disinfectant residual in the Non-UV AR was higher than the Pre-UV AR, the Non-UV AR yielded higher counts (Figure 4.31). As a result, perhaps UV treated water had a negative effect on biofilm *E. coli*. Otherwise, the resistance to NH_2Cl disinfection could have been a result of nitrification in the Non-UV treated reactor.

Unlike both the chlorine and chlorine dioxide phases, low chloramines disinfection did not consistently decreased *E. coli* counts to below the detection limit in Post-UV reactors containing cast iron or polycarbonate coupons. Biofilm *E. coli* counts rose throughout high chloramines disinfection. As mentioned in previous sections, a plausible explanation for this phenomenon could have been nitrification within these reactors.

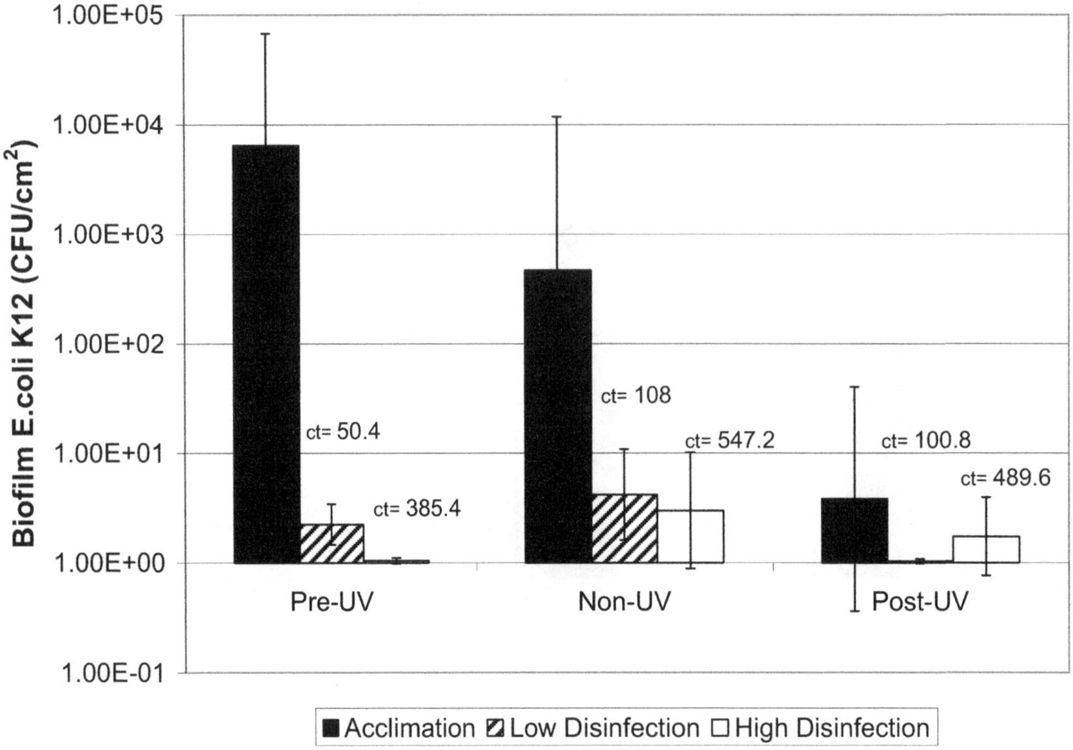

Figure 4.31 Biofilm *E. coli K12* in cast iron ARs during chloramines disinfection

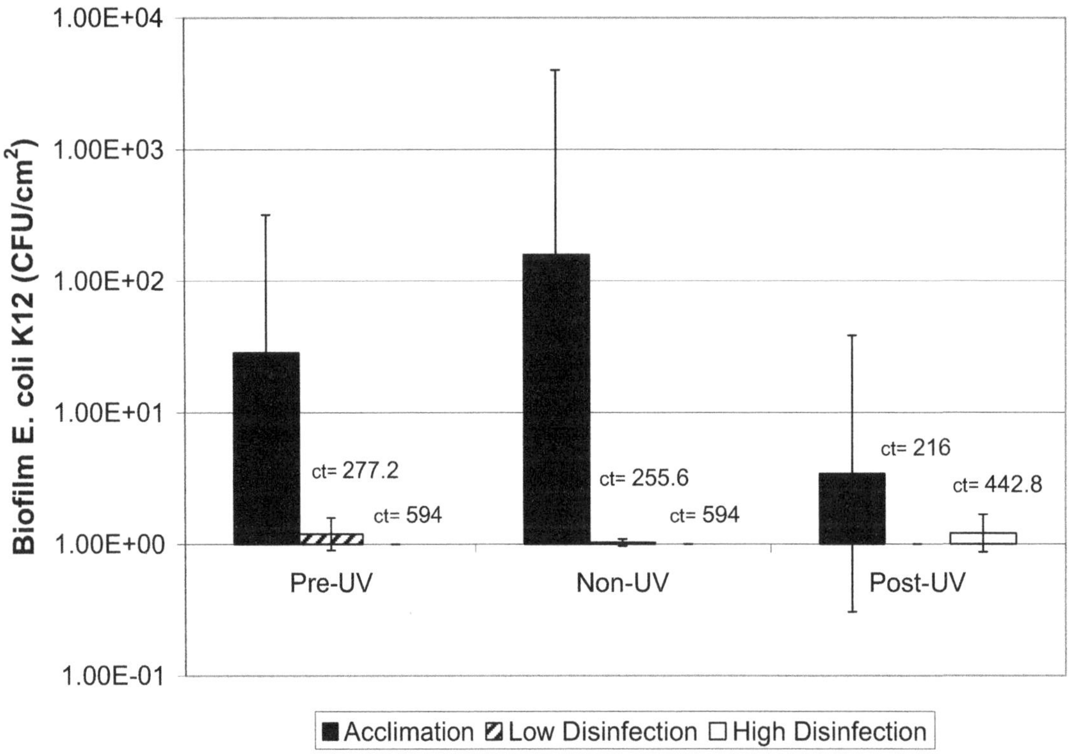

Figure 4.32 Biofilm *E. coli K12* in polycarbonate ARs during chloramines disinfection

Analysis of Biofilm *E. coli*

Overall, for each disinfectant trial there was a noticeable difference among biofilm *E. coli* that was Post –UV treated versus *E. coli* that was not UV irradiated. In most cases these results were significant (Table 4.15). In general, there was no significant distinction between Pre and Non-UV disinfection during acclimation. Table 4.15 displays the ANOVA results when comparing the three UV groups. The shaded sections of the table indicate where the treatments were significantly dissimilar.

In summary, no consistent conclusions can be drawn concerning the impact of UV disinfection (Pre-UV) on the ability of *E. coli* to attach to the biofilm of cast iron or polycarbonate surfaces during a simulated cross contamination intrusion. Pozos *et al* (2004) drew similar conclusions during a study that introduced two opportunistic pathogens (*Escherichia coli* EC98 and *Klebsiella pneumoniae* KP02), into ARs that received UV irradiated influent water.

Effect of Pipe Material

In general, initial biofilm *E. coli* during acclimation in annular reactors containing cast iron did not vary considerably from those containing polycarbonate coupons (Table 4.16). There were two cases in which case iron surfaces supported a higher amount of biofilm *E. coli* than polycarbonate surfaces.

In contrast, there were several significant trends found during the disinfection phase. It appears as though the *E. coli* mutant strain on cast iron surfaces was more resistant to chlorine disinfection than mutants on polycarbonate surfaces (Table 4.16). Also, *E. coli K12* was more resistant to chlorine dioxide disinfection on cast iron surfaces than polycarbonate (Figure 4.29-4.30; Table 4.16). In the chloramines trial, the Non-UV reactors containing *E. coli K12* had higher counts on cast iron coupons than polycarbonate (Figures 4.31-4.32). This was also the case for the Pre-UV reactors; however this result was not significant (Table 4.16).

.

Table 4.16
P-values for the difference between cast iron and polycarbonate coupons on biofilm *E. coli* counts

Phase	Disinfectant	*E. coli K12*			*E. coli* Mutant	
		Pre- UV	Non-UV	Post-UV	Pre- UV	Non-UV
Acclimation	Cl_2	0.649	0.867	0.707	0.928	0.709
	ClO_2	0.208	0.999	0.299	0.607	0.769
	NH_2Cl	0.020	0.659	0.951	0.958	0.027
Disinfection	Cl_2	0.885	0.424	0.223	0.001	0.006
	ClO_2	0.001	0.002	0.152	0.154	0.118
	NH_2Cl	0.064	0.003	0.406	0.087	0.216

Effect of Secondary disinfectants

Figures 4.33 and 4.34 illustrate the average changes in biofilm *E. coli K12* from acclimation, low disinfection, and high disinfection for Cl_2, ClO_2 and NH_2Cl in cast iron and polycarbonate systems.

Cast Iron. Biofilm *E. coli* reduction on cast iron surfaces varied significantly from one disinfectant to another ($p = 6.49 \times 10^{-5}$). In the Pre-UV reactors, Pre-UV-Cl_2 followed by Pre-UV- NH_2Cl and Pre-UV-ClO_2 attained the highest to lowest level of *E. coli* reduction during low disinfection. Log drops for these reactors were 3.53 log, 3.46 log, and 2.52 log, respectively (Table 4.12). The same trend occurred in Non-UV treated reactors ($p=2.0 \times 10^{-4}$) where log decreases in low disinfection were 3.27, 2.05 and 0.28, for Cl_2, NH_2Cl and ClO_2, in that order (Table 4.12). The extraordinarily low *E. coli* reduction in the ClO_2 reactor may have been the result of an increased disinfectant demand because of iron oxidation and iron sloughing off the surface. The turbidity in that reactor during low disinfection reached as high as 10-20 NTU at various points.

For Post-UV treated systems, the trends differed. There were no significant variations among the chemical disinfectants. Log reductions varied from 0.55- 1.39 log. ClO_2 achieved the greatest log reduction; however this was probably the case because it also had the highest initial amount of biofilm *E. coli* (Figure 4.33). All three disinfectants in the Post-UV ARs reduced biofilm *E. coli* to 0-1 CFU/mL after low disinfection. Interestingly, *E. coli* counts increased in the Post-UV-NH_2Cl AR during high disinfection. Once again, this is likely due to nitrification within the AR. Another probable explanation could be that after UV and low disinfection, the *E. coli* colonies were injured and not able to be detected on the selective m-Endo media used in these experiments (McFeters *et al*, 1986a). Therefore during low disinfection where the colonies are non- detect, perhaps they were actually present but could only be cultured during high disinfection (Figure 4.33).

Polycarbonate. In contrast to the cast iron ARs, polycarbonate ARs had consistent biofilm *E. coli K12* log reductions between chemical disinfectants in the Pre-UV, Non-UV and Post-UV reactors during low and high disinfection. Log decreases ranged from 0.09 to 4.05log during low disinfection (Table 4.16, Figure 4.34). In Figure 4.34, both Pre-UV Cl_2 and ClO_2 reactors maintained the same Ct of 93.6. At this equivalent Ct, chlorine dioxide slightly outperformed chlorine with *E. coli* decreases of 4.05 log and 3.89 log, respectively. In the Non-UV treated reactors, Cl_2 (3.54 log) attained a slightly higher coliform log reduction over ClO_2 (2.84 log). Overall, it was determined that chlorine, chlorine dioxide and chloramines were statistically equivalent on the basis of biofilm *E. coli* reduction on polycarbonate surfaces.

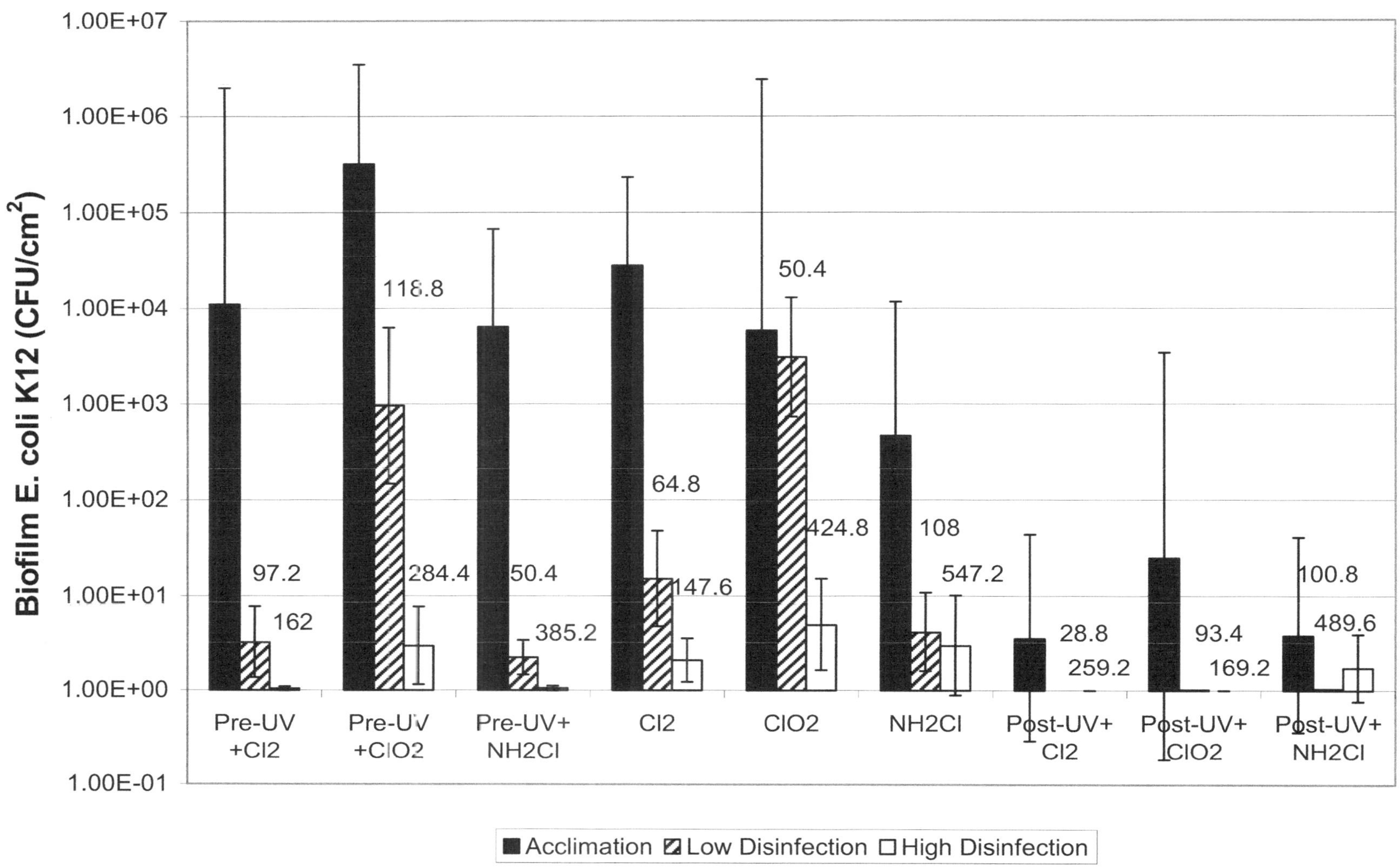

Figure 4.33 Biofilm *E. coli K12* in cast iron ARs

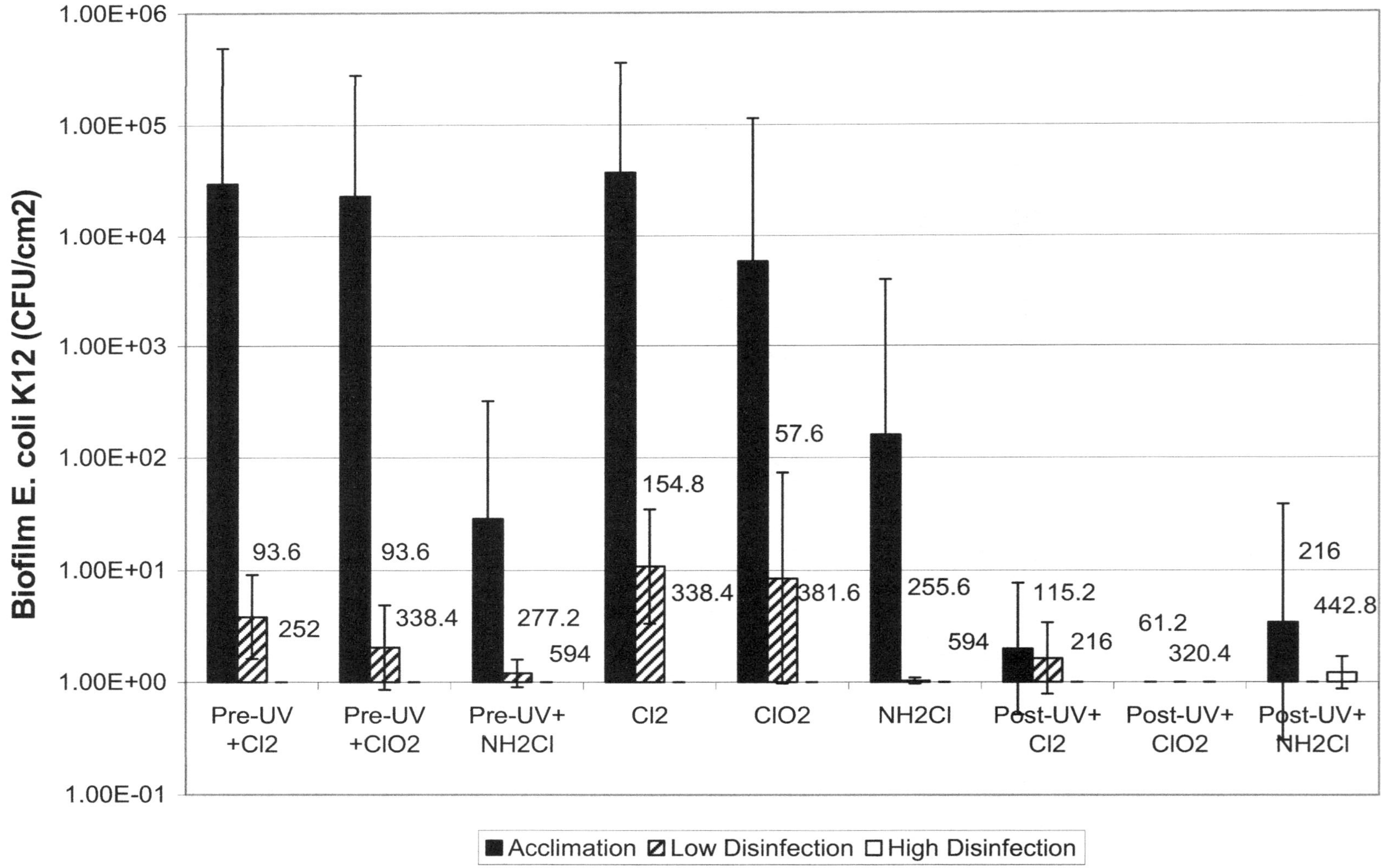

Figure 4.34 Biofilm *E. coli K12* in polycarbonate ARs

Coliform Ecology-Phenotyping

A phenotype is defined as any characteristic of an organism that can be detected in terms of appearance, structure or some measurable property (Mathews *et al*, 1999). Phenotype bioassays are used to identify and classify organisms based on their phenotypic responses to various substrates. In this study, the API 20E identification system was used to assess if phenotypic changes occurred in the *E. coli* colonies isolated from the biofilm of the ARs during the acclimation and low disinfection phases. Some colonies were isolated from the effluent, however no phenotypical changes were ever observed.

Table 4.17
API 20E Substrate identification table

Tests	Substrate	Reaction / Enzymes
ONPG	ONPG	beta-galactosidase
ADH	arginine	arginine dihydrolase
LDC	lysine	lysine decarboxylase
ODC	ornithine	ornithine decarboxylase
CIT	citrate	citrate utilization
H2S	sodium thiosulfate	H2S production
URE	urea	urea hydrolysis
TDA	tryptophane	deaminase
IND	tryptophane	indole production
VP	sodium pyruvate	acetoin production
GEL	gelatin	gelatinase
GLU	glucose	fermentation/oxidation
MAN	mannitol	fermentation/oxidation
INO	inositol	fermentation/oxidation
SOR	sorbitol	fermentation/oxidation
RHA	rhamnose	fermentation/oxidation
SAC	sucrose	fermentation/oxidation
MEL	melibiose	fermentation/oxidation
AMY	amygdalin	fermentation/oxidation
ARA	arabinose	fermentation/oxidation

**Adapted from: (bioMerieux, Inc, 2005).*

The API 20E test kits consist of 20 microtubes containing dehydrated substrates (Table 4.17).These microtubes were inoculated with *E. coli* colonies isolated from the AR biofilms and then incubated. During incubation, color changes triggered by metabolic activity occurred spontaneously or by the addition of reagents. Once all reactions were complete, the strip results (positive/ negative reactions) were recorded and then examined to see if there were any phenotypic trends between pipe materials, disinfectants and/or UV treatment. The positive and negative reactions for all disinfectant trials for both biofilm *E. coli* strains are presented in Tables 4.18-4.23. Highlighted reactions in these tables indicate where the phenotypic response differed from the original stock that was spiked into the reactors.

Chlorine Trial

Tables 4.18 and 4.19 display the results from the chlorine trial for *E. coli K12* and the mutant strain. In general, there were no evident trends that distinguished one strain from the other. Both strains had identical stock profiles and experienced similar changes during chlorine disinfection.

The most significant substrate that varied from the stock was the ONPG substrate. In seven of the eight reactors, it was found that ONPG was not used by the isolated *E. coli* colonies during chlorine disinfection. Interestingly, ONPG is used as an indicator in industry-accepted substrate based coliform detection systems such as Coliert® that was used in this study to detect coliforms in the AR effluent (see Chapter 2). Since the coliforms isolated from the ARs' biofilms did not use the ONPG substrate, it is plausible that false negatives occurred in the effluent of the ARs during the Coliert® tests. Coliforms may have been present, but unable to produce the necessary enzymatic response to process the indicator substrates. Although these results are preliminary, this finding is important because the majority of public utilities in North America use chlorine as a disinfectant and virtually all of them use Coliert® or a similar product to detect coliforms in their treated and distributed water (Trussell 1999, Olstadt et al. 2005). Therefore, the potential for false negatives in chlorinated distributed water is an important public health concern. False negative occurrences as a result of sub-lethal injury to coliforms have been documented by other researchers using industry accepted detection methods such as Coliert® (McFeters et al. 1993, Olstadt et al. 2005).

Two other substrates that differed from the stock during disinfection were glucose (GLU) and rhamnose (RHA). Both substrates were not used by *E. coli* enzymes during low disinfection. Consequently, perhaps in response to stress or injury, the *E. coli* no longer oxidized glucose or rhamnose. Results varied from one AR to the next with no evident trends. Arabinose (ARA) was another substrate that was utilized differently than the stock. In this case, arabinose was not oxidized by colonies isolated from cast iron biofilms. Lastly, there were noteable variations between the stock and *E. coli* isolates for the following substrates: sucrose (SAC) and melibiose (MEL).

Table 4.18

API 20E test kit results for *E. coli K12* during the chlorine trial

Substrate	Stock	Acclimation				Low Cl_2 Disinfection			
	K12*	UV-CI*	UV-PC*	CI*	PC*	UV-CI*	UV-PC	CI	PC*
ONPG	+	+	+	+	+	+	-	-	-
ADH	-	-	-	-	+	-	-	-	-
LDC	+	+	+	+	+	+	+	+	+
ODC	-	-	-	-	-	-	-	-	-
CIT	-	-	-	-	-	-	-	-	-
H2S	-	-	-	-	-	-	-	-	-
URE	-	-	-	-	-	-	-	-	-
TDA	-	-	-	-	-	-	-	-	-
IND	+	+	+	+	+	+	+	+/-	+
VP	-	-	-	-	-	-	-	-	-
GEL	-	-	-	-	-	-	-	-	-
GLU	+	-	+	+	+	+	-/+	-	+
MAN	+	+	+	+	+	+	+	+/-	+
INO	-	-	-	-	-	-	-	-	-
SOR	+	+	+	+	+	+	+	+/-	+
RHA	+	-	+	+	+	-	+/-	+	+
SAC	-	-	-	-	-	-	-	-	-
MEL	+	+	+	+	+	+	-/+	-/+	+
AMY	-	-	-	-	-	-	-	-	-
ARA	+	-	+	+	+	-	+	+/-	+

* *Duplicates*

Table 4.19

API 20E test kit results for *E. coli MMR* mutant during the chlorine trial

Substrate	Stock	Acclimation				Low Cl_2 Disinfection			
	Mut*	UV-CI*	UV-PC*	CI*	PC*	UV-CI	UV-PC*	CI	PC
ONPG	+	+	+	+	+	-	-	+/-	nd
ADH	-	-	+	-	+	-	-	-	nd
LDC	+	+	+	+	-	+	+	+	nd
ODC	-	-	-	-	+	-	-	-	nd
CIT	-	-	-	-	+	-	-	-	nd
H2S	-	-	-	-	-	-	-	-	nd
URE	-	-	-	-	-	-	-	-	nd
TDA	-	-	-	-	-	-	-	-	nd
IND	+	+	+	+	-	+	+	+	nd
VP	-	-	-	-	+	-	-	-	nd
GEL	-	-	-	-	-	-	-	-	nd
GLU	+	+	+	+	+	+	-	+	nd
MAN	+	+	+	+	+	+	-	+	nd
INO	-	-	-	-	-	-	-	-	nd
SOR	+	+	+	+	+	+/-	-	+	nd
RHA	+	+	+	+	-	-	+	-/+	nd
SAC	-	-	-	-	+	-	-	-	nd
MEL	+	+	+	+	+	-/+	+	+	nd
AMY	-	-	-	-	+	-	-	-	nd
ARA	+	+	+	+	+	-/+	+	-/+	nd

* *Duplicates*

nd= non- detect

Chlorine Dioxide Trial

Tables 4.20 and 4.21 display the results from the chlorine dioxide trial. Similar to the chlorine trial, there were no evident differences between *E. coli K12* and the mutant strain with respect to phenotypical responses. There were two significant substrates that differed from the stock in all colonies isolated from the reactors: arginine (ADH) and tryptophane (TDA). Strangely, they are not the same as those from the chlorine trial. The following three substrates were utilized by enzymes in colonies from polycarbonate biofilms only: citrate (CIT), sodium pyruvate (VP) and sucrose (SAC). The isolates appeared to be adapting to their environment by using alternative energy sources to survive in the polycarbonate biofilms. From the results, it seems as though citrate and sucrose were utilized as energy sources and sodium pyruvate was consumed for acetoin production (*bioMerieux Inc, 2005).* Fermentation/ oxidation of melibiose (MEL) was exhibited by *E. coli* taken from Pre-UV-PC biofilms (K-12 & Mutant). Finally, amygdalin (AMY) oxidation by the enzyme amylase occurred on numerous occasions in isolates from both cast iron and polycarbonate surfaces (Tables 4.24 and 4.25).

Table 4.20

API 20E test kit results for *E. coli K12* during the chlorine dioxide trial

Substrate	Stock	Acclimation				Low ClO_2 Disinfection			
	K12*	UV-CI*	UV-PC*	CI*	PC*	UV-CI*	UV-PC	CI	PC
ONPG	+	+	+	+	+	+	+	+	+
ADH	-	+	+	+	+	+	+	+	+
LDC	+	+	+	+	+	+	+	+	+
ODC	-	-	-	-	-	-	-	-	-
CIT	-	-	-	-	-	-	+	-	-/+
H2S	-	-	-	-	-	-	-	-	-
URE	-	-	-	-	-	-	-	-	-
TDA	-	-	-	-	-	+	+	+	+
IND	+	+	+	+	+	+	+	+	+
VP	-	-	-	-	-	-	+	-	-/+
GEL	-	-	-	-	-	-	-	-	-
GLU	+	+	+	+	+	+	+	+	+
MAN	+	+	+	+	+	+	+	+	+
INO	-	-	-	-	-	-	-	-	-
SOR	+	+	+	+	+	+	+	+	+
RHA	+	+	+	+	+	+	+	+	+
SAC	-	-	-	-	-	-	+	-	-/+
MEL	+	+	+	+	+	+	+/-	+	+
AMY	-	-	-	-	-	+	+	-/+	-/+
ARA	+	+	+	+	+	+	+	+	+

** Duplicates*

Table 4.21
API 20E test kit results for *E. coli* MMR mutant during the chlorine dioxide trial

Substrate	Stock	Acclimation				Low ClO_2 Disinfection			
	Mut*	UV-CI*	UV-PC*	CI*	PC*	UV-CI	UV-PC*	CI*	PC*
ONPG	+	+	+	+	+	+	+	+	+
ADH	-	+	+	+	+	+	+	+	+
LDC	+	+	+	+	+	+	+	+	+
ODC	-	-	-	-	-	-	-	-	-
CIT	-	-	-	-	-	-	+	-	-
H2S	-	-	-	-	-	-	-	-	-
URE	-	-	-	-	-	-	-	-	-
TDA	-	-	-	-	-	+	+	+	+
IND	+	+	+	+	+	+	+	+	+
VP	-	-	-	-	-	-	+	-	-
GEL	-	-	-	-	-	-	-	-	-
GLU	+	+	+	+	+	+	+	+	+
MAN	+	+	+	+	+	+	+	+	+
INO	-	-	-	-	-	-	-	-	-
SOR	+	+	+	+	+	+	+	+	+
RHA	+	+	+	+	+	+	+	+	+
SAC	-	-	-	-	-	-	+	-	-
MEL	+	+	+	+	+	+	-	+	+
AMY	-	-	-	-	-	-/+	+	-	-
ARA	+	+	+	+	+	+	+	+	+

** Duplicates*

Chloramine Trial

Tables 4.22 and 4.23 present the results from the chloramines trial. Substrates that were utilized by coliforms during disinfection and not in the stock included: arginine (ADH), citrate (CIT), ornithine (ODC), sodium pyruvate (VP), gelatin (GEL), amygladin (AMY), and sucrose (SAC). The following substrates were not used by enzymes during monochloramine disinfection: arabinose (ARA), lysine (LDC), tryptophane (TDA), mannitol (MAN), sorbitol (SOR), and melibiose (MEL). The tryptophane substrate differed only from the stock in *E. coli* isolated from cast iron ARs. The others varied between the two pipe materials.

In cases where lysine (LDC) was not consumed, the enzyme lysine decarboxylase may not have been functioning because the *E. coli* was under acid stress conditions. When *E. coli* is under acid stress conditions, lysine and ornithine decarboxylase may not function because they provide acid stress protection (Storz *et al*, 2000). Acid was used to neutralize the high pH of the monochloramine dosed into the ARs. Although the pH of the AR effluent remained fairly constant throughout disinfection (pH= 7.4), perhaps the addition of acid stressed *E. coli* within the AR biofilms.

Table 4.22

API 20E test kit results for *E. coli K12* during the chloramines trial

Substrate	Stock	Acclimation				Low NH_2Cl Disinfection			
	K12*	UV-CI*	UV-PC*	CI*	PC*	UV-CI	UV-PC*	CI	PC
ONPG	+	+	+	+	+	+	+	+	+
ADH	-	-	-	-	-	+	-	+/-	-/+
LDC	+	+	+	+	+	+	+	-	+
ODC	-	-	-	-	-	-	-	-	-/+
CIT	-	-	-	-	-	-	-	+	+/-
H2S	-	-	-	-	-	-	-	-	-
URE	-	-	-	-	-	-	-	-	-
TDA	-	-	-	-	-	+	-	+	-
IND	+	+	+	+	+	+	+	+	+
VP	-	+	+	+	+	-	-	+	-
GEL	-	-	-	-	-	+	-	+	-
GLU	+	+	+	+	+	+	+	+	+
MAN	+	+	+	+	+	+	+	-	+
INO	-	-	-	-	-	-	-	-	-
SOR	+	+	+	+	+	-/+	+	-	+
RHA	+	+	+	+	+	+	+	+	+
SAC	-	-	-	-	-	-	-	-	-
MEL	+	+	+	+	+	+	+	+/-	-/+
AMY	-	-	-	-	-	-	-	+	-
ARA	+	+	+	+	+	-/+	+	+	+

** Duplicates*

Table 4.23

API 20E test kit results for *E. coli MMR* mutant during the chloramines trial

Substrate	Stock	Acclimation				Low NH_2Cl Disinfection			
	Mut*	UV-CI*	UV-PC*	CI*	PC*	UV-CI	UV-PC*	CI*	PC
ONPG	+	+	+	+	+	+	+	+	+
ADH	-	-	-	-	-	-/+	-	+	-
LDC	+	+	+	+	+	+	-	+	-
ODC	-	-	-	-	-	+/-	-	+	-
CIT	-	-	-	-	-	-	+	-	+
H2S	-	-	-	-	-	-	-	-	-
URE	-	-	-	-	-	-	-	-	-
TDA	-	-	-	-	-	+	-	-	-
IND	+	+	+	+	+	-	+	+	-
VP	-	+	+	+	+	+	-	-	+
GEL	-	-	-	-	-	+	-	-	+
GLU	+	+	+	+	+	+	+	+	-
MAN	+	+	+	+	+	+	+	+	-
INO	-	-	-	-	-	-	-	-	-
SOR	+	+	+	+	+	+/-	+	+	-
RHA	+	+	+	+	+	+	+	+	+
SAC	-	-	-	-	-	-	+	-	-
MEL	+	+	+	+	+	-/+	-	+	+
AMY	-	-	-	-	-	-	+	-	+
ARA	+	+	+	+	+	+	+	+	+

** Duplicates*

DGGE Analysis

To date, 114 individual bands have been excised. Of those to date, 87 have been sequenced. Those 87 break down as follows: Chlorine run: 12 sequenced; Chlorine Dioxide run: 31 sequenced; Chloramine run: 30 sequenced; RAW water: 5 sequenced; GAC water, 4 sequenced; UV water: 2 sequenced; Marsha's white flakes: 3 sequenced.

The data is summarised in Tables 4.24-4.27 for the strains identified to date. These are accompanied by representative DGGE gels, one pair for each disinfectant used (Figures 4.35 – 4.40) showing the bands obtained without and with treatment at the lower concentration tested. Differences between substrates (cast iron and poly carbonate) as well as differences occurring with and without upstream UV treatment can be seen on the same gels.

Tables 4.24-4.27 are inevitably complex to summarize all the data for each disinfectant together. They delineate the bacteria identified in the bands excised so far, and where similar bands are observed in other gels. The criteria used for excision of gel bands were so far primarily the observation of clear similarities or differences between gels that used similar experimental conditions. The legend for the tables below is shown briefly on each Table but deserves a little more explanation here, as follows:

- B = Biofilm isolation: DNA extracted from material recovered from AR coupons
- E = Effluent isolation: DNA extracted from aliquot of effluent from AR
- C(B) = Cultured Biofilm isolation: bacteria recovered from biofilm by culture before DNA extraction
- C(E) = Cultured Effluent isolation: bacteria recovered from effluent by culture before DNA extraction
- \+ = UV pre-treatment present: upstream UV treatment was in place
- \- = UV pre-treatment absent: no upstream UV treatment
- **Bold** = bold values indicate the operational conditions for the gel from which the band was excised
- Non-Bold = non-bold values show other experimental conditions from which the band was identified by comparison with other lanes and DGGE gels.
- Blast and Ribosomal Database Project II (RDP) were used to identify the bands: values are given to provide a reference of how similar the sequence obtained is to the database value. The highest similarity value for a known species obtained from both databases was chosen as the identity of the excised band. A value of less then 90% for blast, and 85% for RDP, was set as our arbitrary cut-off for assigning an identification. These numbers are set at different values in different studies, sometimes 95% and identification is often not completely definitive, however, we wanted to try and assign as many tentative identities as possible. Unknown species and uncultured species were only chosen when the databases did not turn up a known species above our arbitrary cut-off value. In some cases, due to the short nature of the sequences obtained, we could only identify the band down to the Genus level.

To illustrate how the tables are read we have chosen the example of *Pseudomonas lanceolata* in the chloramine experiment, formatted slightly differently than the Table. As indicated here, this organism was found on both substrates and in both untreated and low chloramine concentration but no band was seen at 1.0 mg/L chlorine. It was seen by DGGE in the biofilm with and without UV (B+, B-) and the effluent with and without UV (E+, E-), and also in culture for both the biofilm and effluent C(B-) & C(E+) in the untreated samples but once treatment commenced it was only found in the biofilm samples whether or not they had been pretreated with UV and whether or not the DNA was extracted directly from the biofilm or after prior culture on R2A agar.

Table 4.24

Identification and Comparison of Bands by PCR-DGGE for cast iron and polycarbonate

	0 mg/L	0.2 mg/L	1.0 mg/L
CAST IRON	B+, B-, C(B)-, C(E)+, C(E)-	B+, B-, C(B)+, C(B)-	
POLYCARBONATE	B+, B-, E+, E-, C(B)-, C(E)+, C(E)-	B+, C(B)+, C(B)-	

Several species are observed in all the three chemical treatments as seen in Tables 4.25 – 4.27. Many Pseudomonas species have been identified in all chemical treatments, and are found in almost every possible set of conditions. This is not a surprise since pseudomonads are ubiquitous biofilm organisms. *Aquabacterium parvum*, *Dechloromonas sp.*, *Methylobacillus flagellatus*, *Pseudomonas mendocina*, *Escherichia coli*, *Methylobacillus flagellatum*, and Uncultured *Bacteroidetes* bacterium seem to only be found on cast iron coupons, while *Chitinophaga sp.* was only observed on polycarbonate coupons. Several of the species identified were also observed from culture on R2A media from both the biofilm and effluent samples. There are however some species such as *Acinetobacter lwoffii, Aquabacterium parvum, Methylobacillus flagellatus, Pseudomonas mendocina,* Uncultured *Bacteroidetes* bacterium, and Uncultured *Verrucomicrobia* bacteria, that were not found to be cultured using R2A media. Although differences as well as similarities are clearly identified between substrates, the most marked being the absence of dechloromonas on polycarbonate coupons, the differences could be either general and predictive or specific to the water source and not predictive. Since studies of this kind on drinking water are, as yet, very rare, it is difficult to be definitive though we believe that the presence of *Dechloromonas* will be a general phenomenon in cast iron pipes. While it is obvious that some bands disappear following UV treatment, it is equally clear that other bands appear. Obviously the latter must be relatively resistant to UV treatment, though it is not yet known whether these microorganisms are responding to new nutrients made available by the UV treatment, or whether they are simply better able to flourish in the absence of so much competition.

Identification of bands is still ongoing in order to categorize bands which are still unknown, and in order to make comparison of gels easier. Sections containing more analysis are to be added as the analyses proceed. Due to the experimental constraints that necessitated treating each disinfectant separately, and because it is acknowledged that water quality changes seasonally, comparisons between or among disinfectants are less definitive. Nevertheless, certain speculations can be made and a few conclusions can be drawn. The section speculating on observed differences between disinfectants will be written later as we have more definitive identification of bands on the gels. Currently more bands have been identified for chloramine than the other disinfectants, but implications should be drawn from this.

Table 4.25
Identification and Comparison of Bands by PCR-DGGE for the Chloramine Run.

Sequence Identity	Cast Iron 0 mg/L	Cast Iron 0.2 mg/L	Cast Iron 1.0 mg/L	Polycarbonate 0 mg/L	Polycarbonate 0.2 mg/L	Polycarbonate 1.0 mg/L	Blast %	RDP %
Acinetobacter johnsonii	**B-**, C(E)-	**B+**, B-, E+, E-		B-, C(B)-, C(E)-	**E-**	E+	100	98.4
Acinetobacter lwoffii		**B+**		E+	E+		97	93.4
Acinetobacter sp.		**B-**					99	97.3
Acinetobacter sp. SS-2		**B+**					100	100
Aquabacterium parvum	B+	**E+**, E-					100	100
Chitinophaga sp.						C(B)+, **C(B)-**	97	92.3
Curvibacter gracilis	**B+**, E+, C(E)+			B+, C(E)+		E+	97	91.7
Dechloromonas aromatica RCB	B+, **B-**, E-						98	89.2
Dechloromonas sp. CL24	B+, **B-**, E-, C(E)-						100	100
Dechloromonas sp. PC1	**B+**, E+, C(B)+						100	100
Flexibacter sp.			B+	**E+**, **E-**	C(B)+	E+	99	96.2
Methylobacillus flagellatus	**B-**, **E-**						95	**83**
Pseudomonas fluorescens	B+, B-, C(B)+, C(E)+	B+, B-, **C(B)+**	E+	**B+**, B-, C(B)+, C(E)+	B+, B-	E+, E-	98	93.1
Pseudomonas lanceolata	**B+**, B-, C(B)-, C(E)+, C(E)-	B+, B-, C(B)+, C(B)-		B+, B-, E+, E-, C(B)-, C(E)+, C(E)-	B+, **C(B)+**, C(B)-		99	97.3
Pseudomonas mendocina	**E-**						99	97.3
Pseudomonas putida	**B-**				C(B)+, **C(B)-**		100	100
Pseudomonas saccharophila	B+, **B-**, E+, C(E)+, C(E)-					E+	100	100
Shewanella putrefaciens	**E-**	**B+**, B-, E+, E-			B+, B-		100	100
Stenotrophomonas maltophilia	B+, **B-**		**C(B)-**			C(B)-	99	98.9

Table 4.26
Identification and Comparison of Bands by PCR-DGGE for the Chlorine Dioxide Run.

Sequence Identity	Cast Iron 0 mg/L	Cast Iron 0.2 mg/L	Cast Iron 1.0 mg/L	Polycarbonate 0 mg/L	Polycarbonate 0.2 mg/L	Polycarbonate 1.0 mg/L	Blast %	RDP %
Acidovorax delafieldii	C(B)+, **C(B)-**			**B-**, C(B)-			100	100
Aquabacterium sp.	B+, E+	**E+**		E+, C(E)+, C(E)-	B-		100	100
Dechloromonas sp. CL24	B+, B-, E+, C(B)+, **C(B)-**, C(E)+	**B+**, B-, E+, E-, C(B)+ C(B)-	B+, B-, C(B)+, C(B)-				100	97.3
Dechloromonas sp. PC1	B+, **B-**, E-, C(B)+, **C(B)-**	**B+**, B-, E+, E-, C(B)+ C(B)-	B+, B-, C(B)+, C(B)-				99	96.7
Escherichia coli	B+, **C(B)+**		C(B)+				100	100
Methylobacillus flagellatum	B+, **B-**, E+, E-	**B+**, B-, E+, E-	**B+**, B-				96	88.3
Phyllobacterium sp.	B-	**B+**, **B-**	B-, E+, E-			**E+**, E-	100	100
Pseudomonas fluorescens	C(E)+, C(E)-			**B-**, E+, E-, C(B)+, C(B)-, C(E)+, **C(E)-**			98	92.5
Pseudomonas lanceolata				B+, **B-**, **C(B)+**, **C(B)-**			97	89.8
Pseudomonas putida	C(E)-	**B+**, C(E)-		E+, E-, C(B)+, C(B)-	B+, E+, E-		99	96.8
Pseudomonas saccharophila	B+, B-, **C(B)+**, **C(B)-**	E+, C(B)+, C(E)+, C(E)-	C(B)+, C(B)-	C(B)+, C(B)-			100	100
Pseudomonas sp.	E-	**B-**		**B-**, E+, E-			99	95.9
Uncultured *Acidobacterium*	B+	**B+**, B-, E+, E-	B+, B-				98	92.9
Uncultured *Bacteroidetes bacterium*	B+, **B-**	**B+**, B-, **E+**, E-	B+, B-				100	92.3
Uncultured *Verrucomicrobia bacteria*	B+, **B-**, E-	B+, B-	B+, B-	E+			94	**81.4**

Table 4.27
Identification and Comparison of Bands by PCR-DGGE for the Free Chlorine Run.

Sequence Identity	Cast Iron			Polycarbonate			Blast %	RDP %
	0 mg/L	0.2 mg/L	1.0 mg/L	0 mg/L	0.2 mg/L	1.0 mg/L		
Dechloromonas sp. CL24	B+, B-	B+, B-, C(B)+, C(B)-	B+, B-, C(B)+, C(B)-				100	100
Pseudomonas lanceolata	E+, E+, E-, C(E)+			B-, E+, E-, C(E)+, C(E)-			99	97.3
Pseudomonas plecoglossicida	E+, E-, C(B)+, C(B)-, C(E)-	B+, B-, C(B)+, C(B)-	B+, B-	B-, E+, E-, C(B)+, C(B)-	B+		100	100
Pseudomonas putida	E-, C(B)+, C(B)-, C(E)-	E-, C(E)-	E+, E-	B-, E+, C(B)+, C(B)-, C(E)+, C(E)-	B+, B-, E-	E-, C(B)-, C(E)-	100	100
Uncultured *Bacteroidetes* bacterium		B+, B-	B-				100	92.3

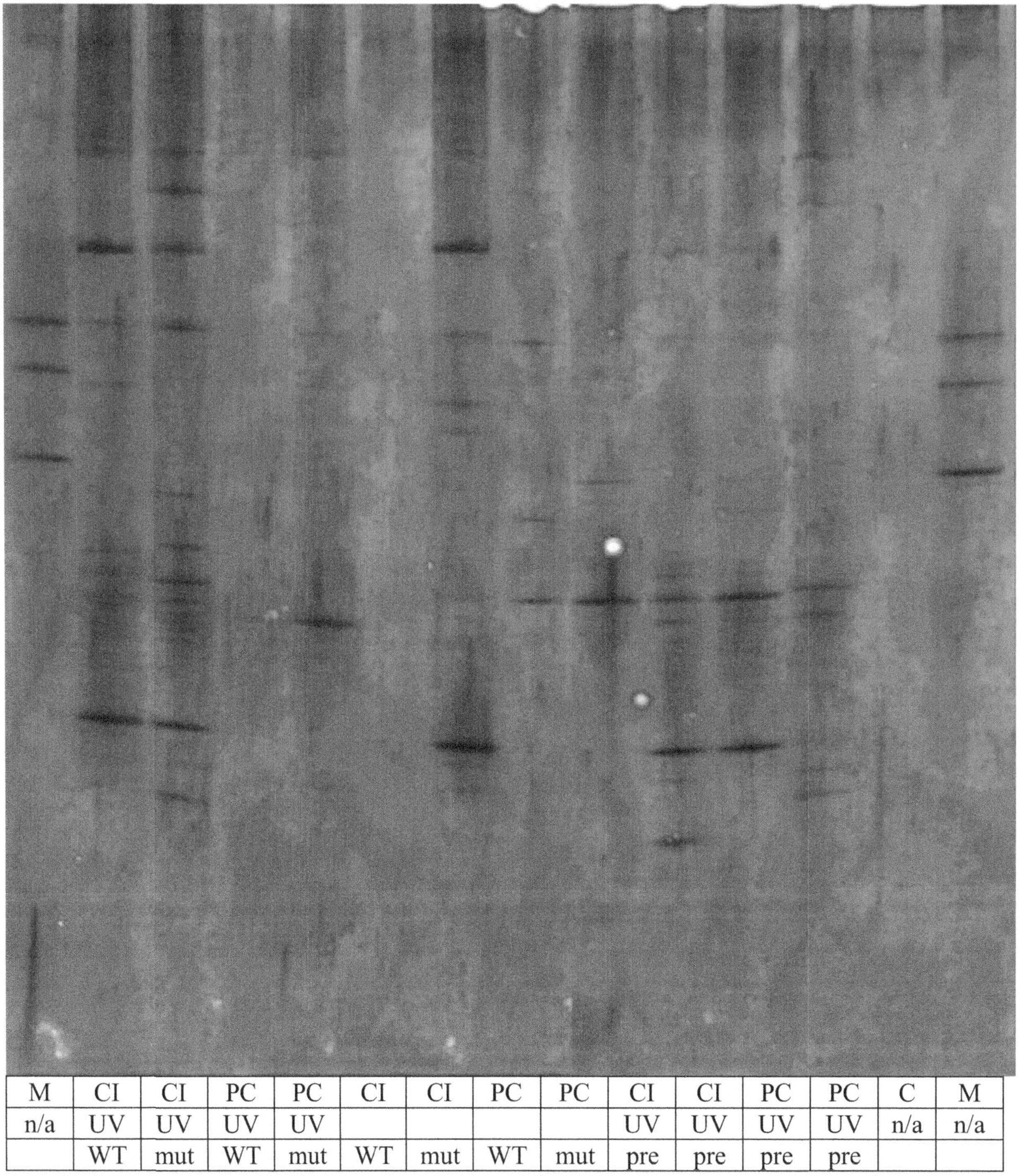

M	CI	CI	PC	PC	CI	CI	PC	PC	CI	CI	PC	PC	C	M
n/a	UV	UV	UV	UV					UV	UV	UV	UV	n/a	n/a
	WT	mut	WT	mut	WT	mut	WT	mut	pre	pre	pre	pre		

M = DGGE marker standard; C = negative control for PCR; CI = cast iron; PC = polycarbonate; WT = wild type *E. coli* spiked; mut = mutant *E. coli* spiked; pre = *E. coli* spiked before UV

Figure 4.35 DGGE gel of PCR products from DNA extracted from the biofilm - pre-treatment phase of the 2nd. Chlorine experiment

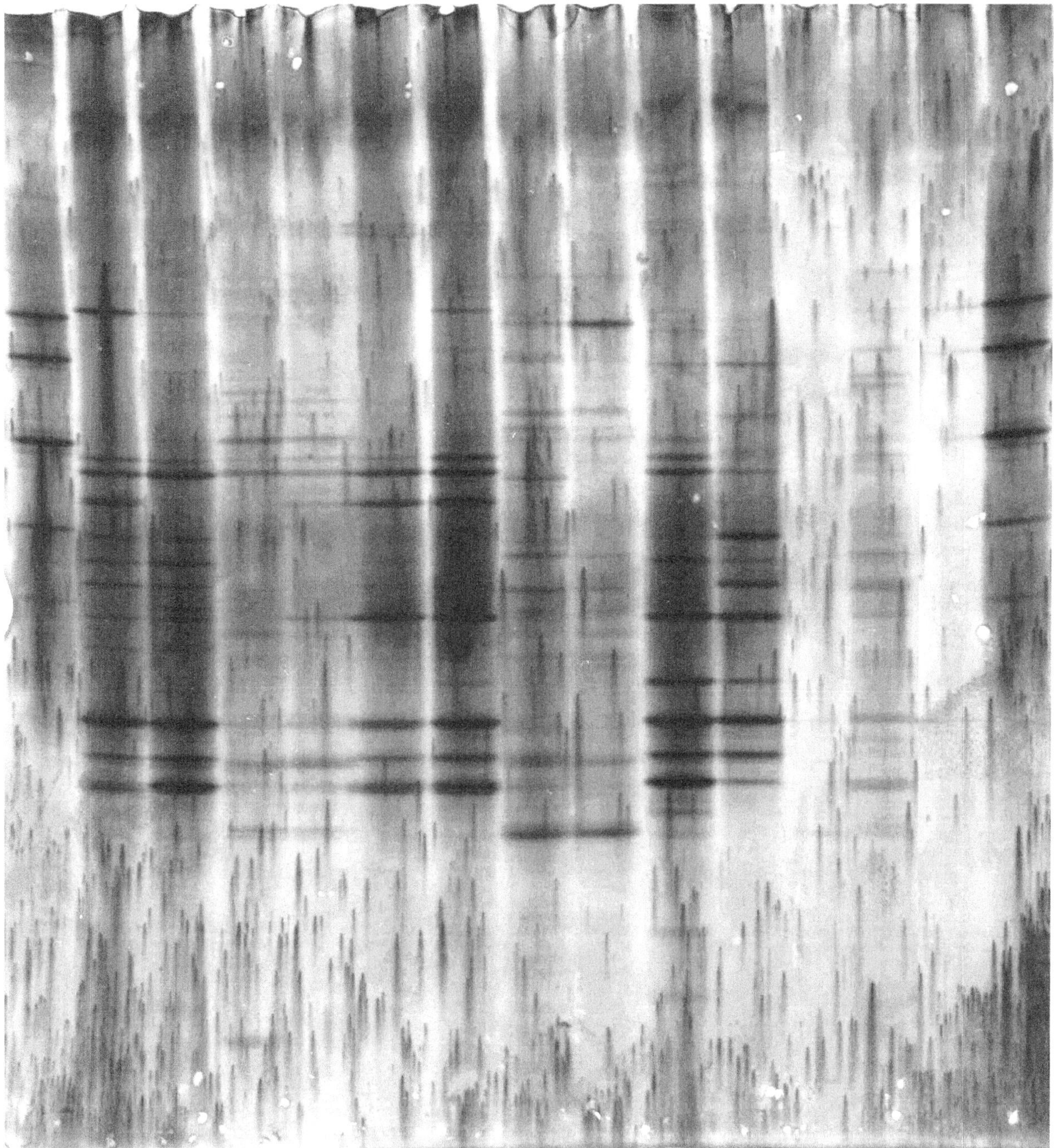

M	CI	CI	PC	PC	CI	CI	PC	PC	CI	CI	PC	PC	C	M
n/a	UV	UV	UV	UV					UV	UV	UV	UV	n/a	n/a
	WT	mut	WT	mut	WT	mut	WT	mut	pre	pre	pre	pre		

M = DGGE marker standard; C = negative control for PCR; CI = cast iron; PC = polycarbonate; WT = wild type *E. coli* spiked; mut = mutant *E. coli* spiked; pre = *E. coli* spiked before UV

Figure 4.36 DGGE gel of PCR products from DNA extracted from the biofilm – low chlorine phase of the 2nd. Chlorine experiment

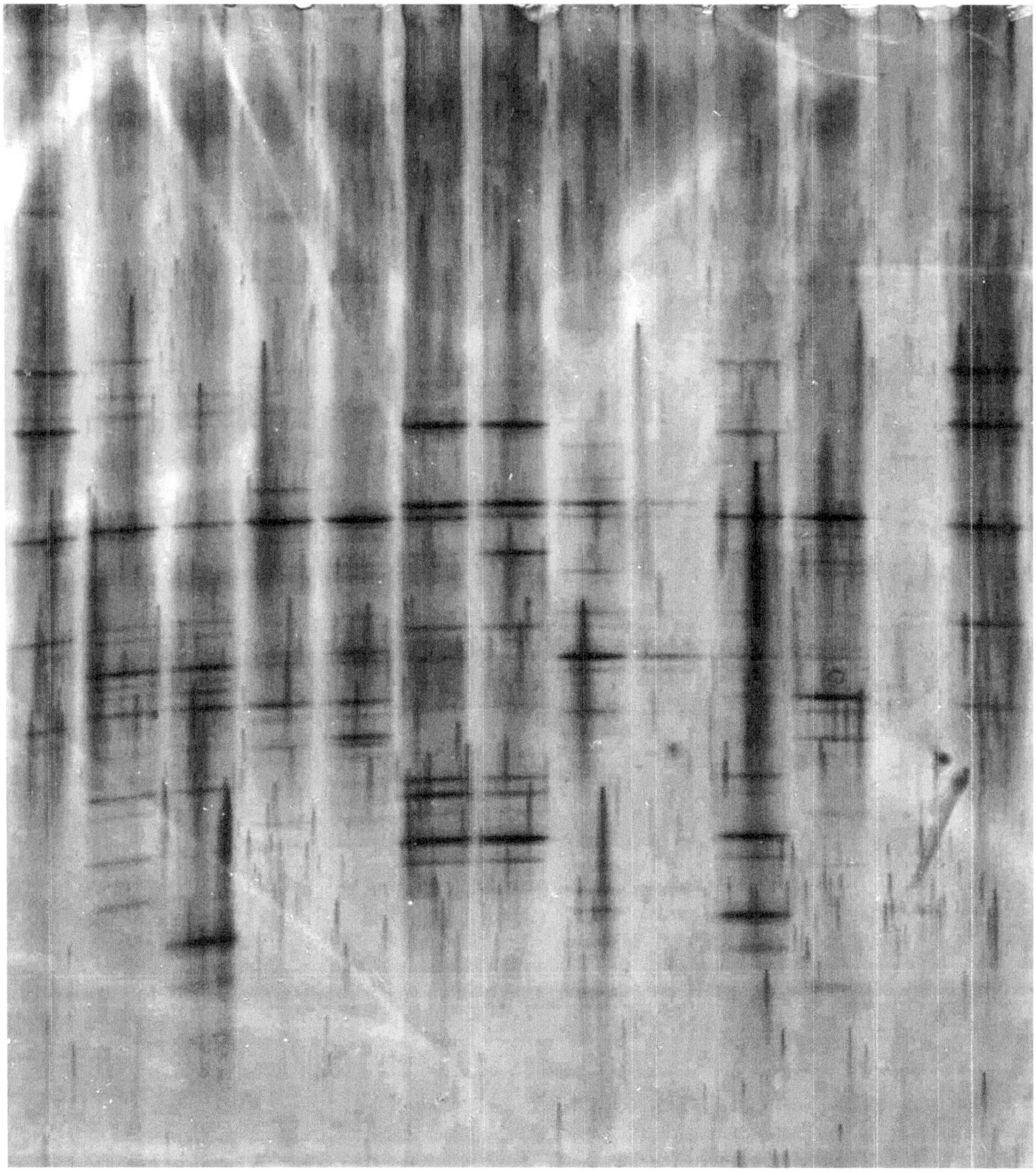

M	CI	CI	PC	PC	CI	CI	PC	PC	CI	PC	C	M
n/a	UV	UV	UV	UV					UV	UV	n/a	n/a
	WT	mut	WT	mut	WT	mut	WT	mut	pre	pre		

M = DGGE marker standard; C = negative control for PCR; CI = cast iron; PC = polycarbonate; WT = wild type *E. coli* spiked; mut = mutant *E. coli* spiked; pre = *E. coli* spiked before UV

Figure 4.37 DGGE gel of PCR products from DNA extracted from the biofilm – pretreatment phase of chloramine experiment

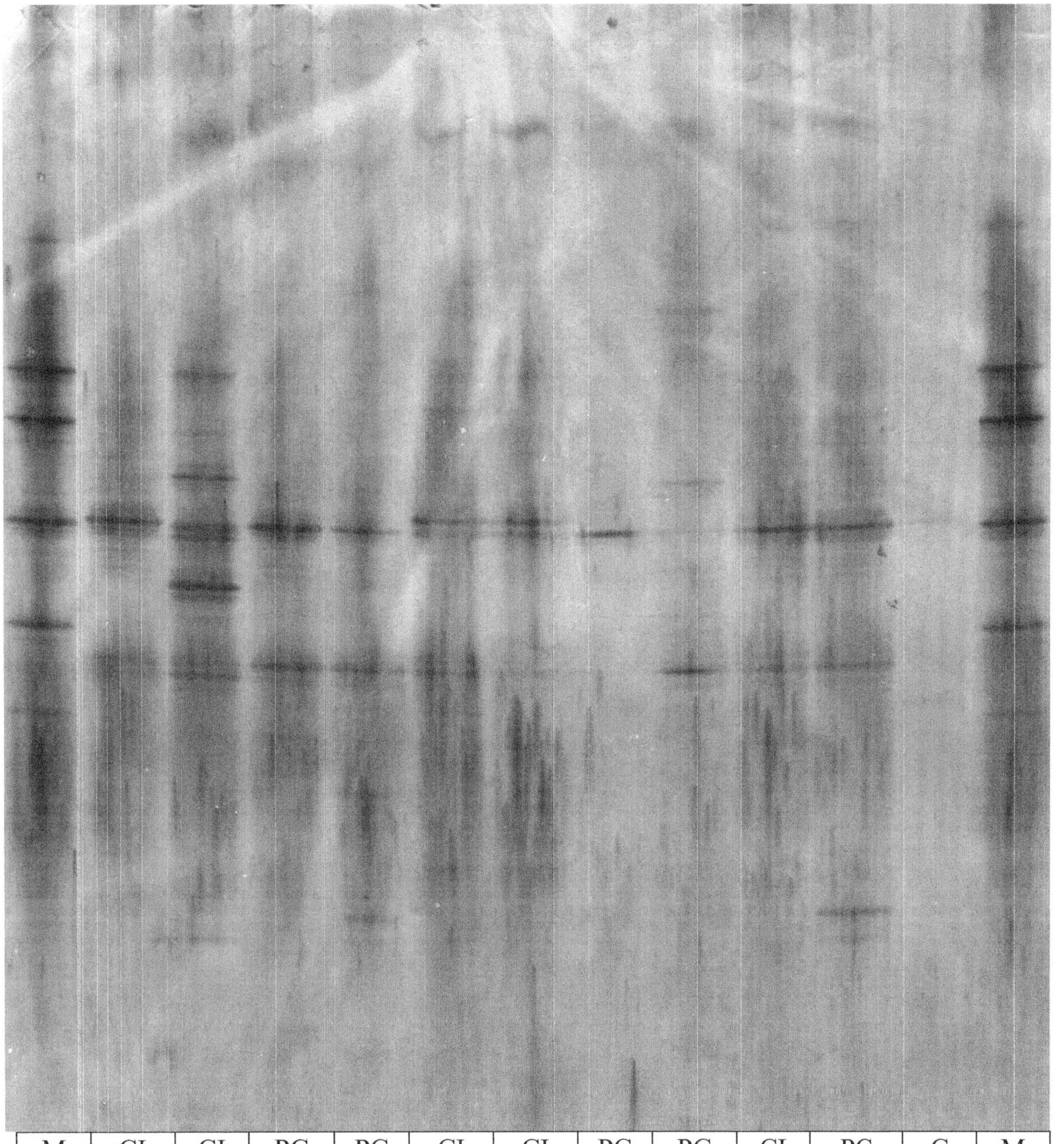

M	CI	CI	PC	PC	CI	CI	PC	PC	CI	PC	C	M
n/a	UV	UV	UV	UV					UV	UV	n/a	n/a
	WT	mut	WT	mut	WT	mut	WT	mut	pre	pre		

M = DGGE marker standard; C = negative control for PCR; CI = cast iron; PC = polycarbonate; WT = wild type *E. coli* spiked; mut = mutant *E. coli* spiked; pre = *E. coli* spiked before UV

Figure 4.38 DGGE gel of PCR products from DNA extracted from the biofilm – low chloramine phase of the chloramine experiment

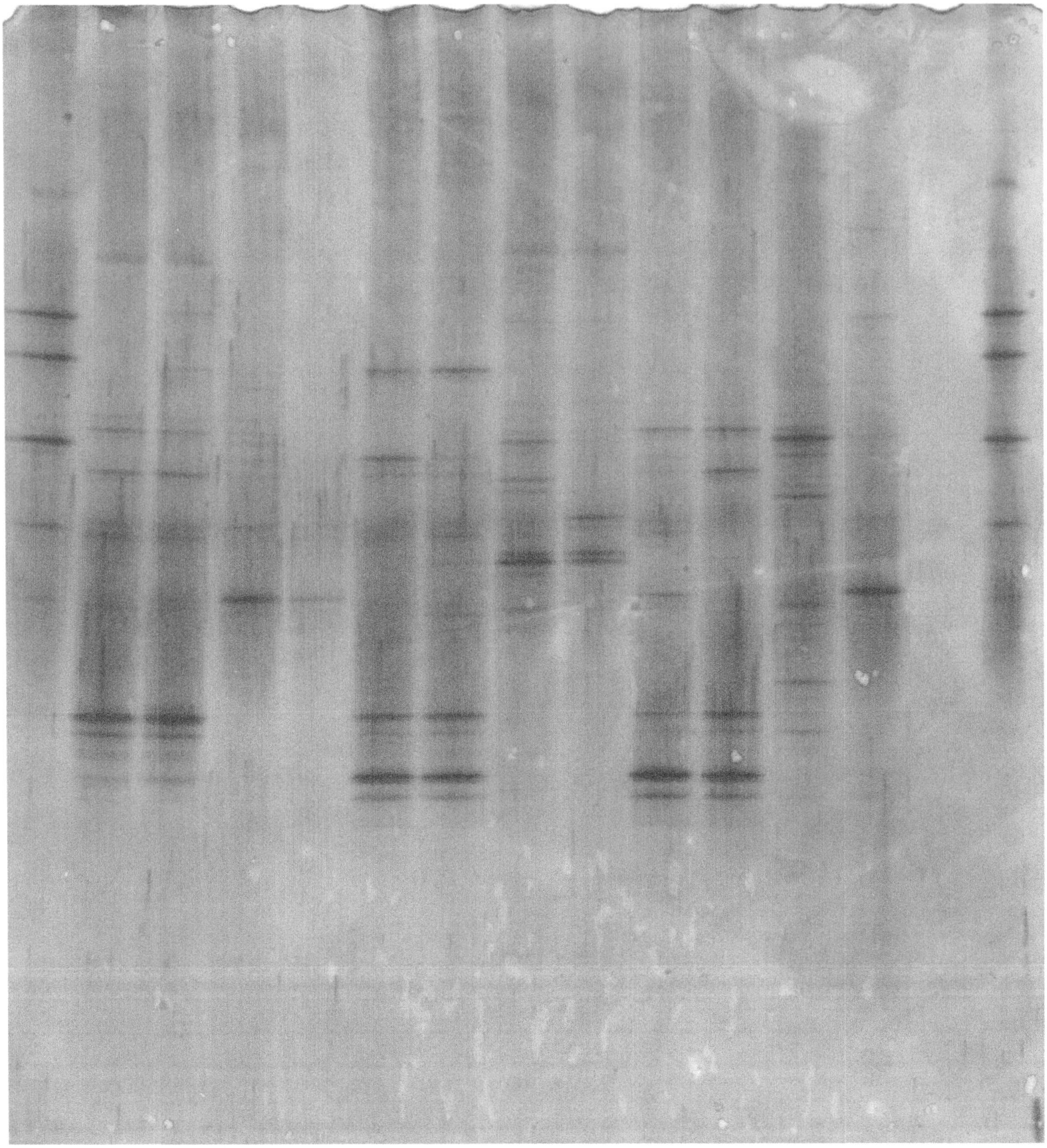

M	CI	CI	PC	PC	CI	CI	PC	PC	CI	CI	PC	PC	C	M
n/a	UV	UV	UV	UV					UV	UV	UV	UV	n/a	n/a
	WT	mut	WT	mut	WT	mut	WT	mut	pre	pre	pre	pre		

M = DGGE marker standard; C = negative control for PCR; CI = cast iron; PC = polycarbonate; WT = wild type *E. coli* spiked; mut = mutant *E. coli* spiked; pre = *E. coli* spiked before UV

Figure 4.39 DGGE gel of PCR products from DNA extracted from the biofilm – pretreatment phase of the chlorine dioxide experiment

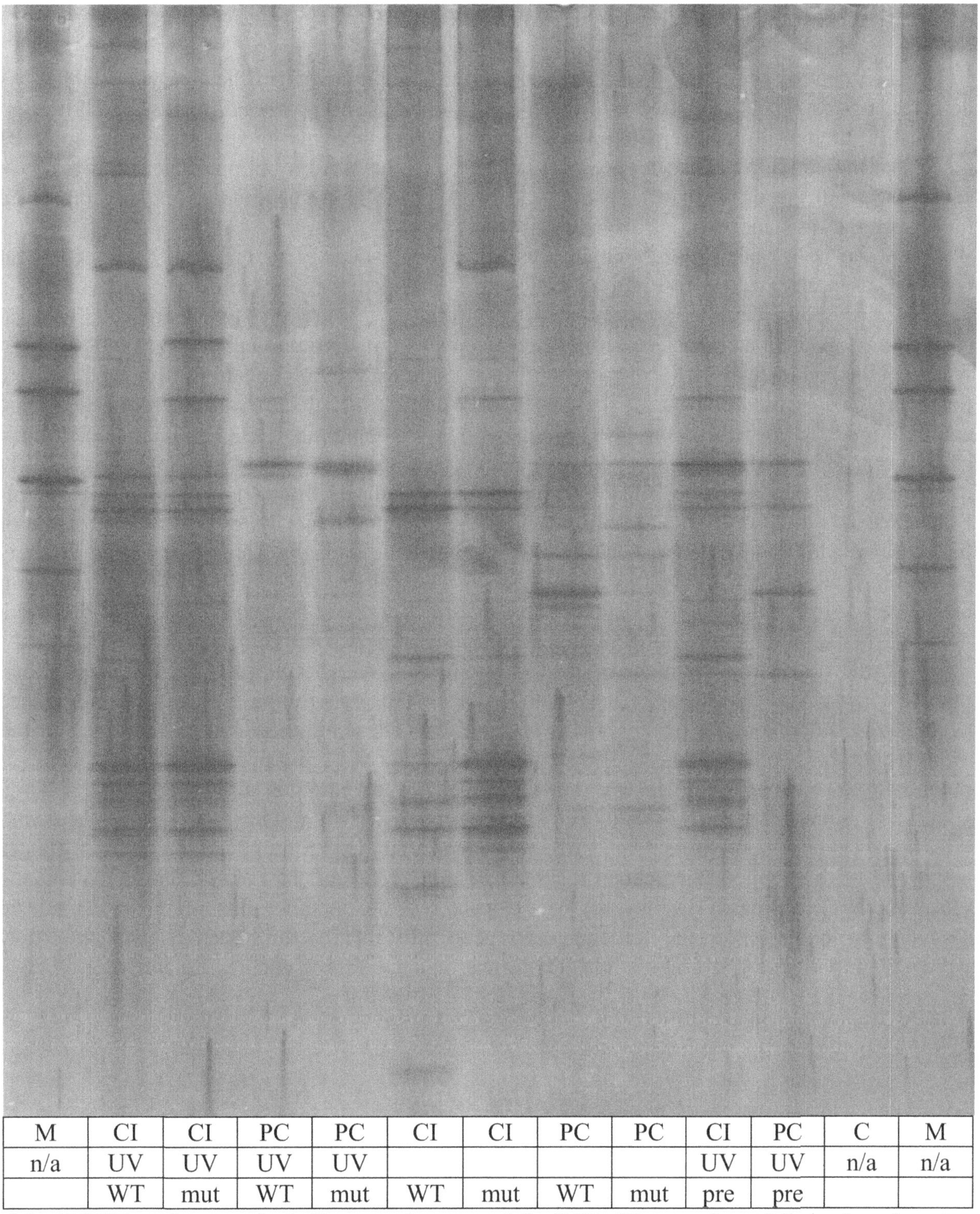

M	CI	CI	PC	PC	CI	CI	PC	PC	CI	PC	C	M
n/a	UV	UV	UV	UV					UV	UV	n/a	n/a
	WT	mut	WT	mut	WT	mut	WT	mut	pre	pre		

M = DGGE marker standard; C = negative control for PCR; CI = cast iron; PC = polycarbonate; WT = wild type *E. coli* spiked; mut = mutant *E. coli* spiked; pre = *E. coli* spiked before UV

Figure 4.40 DGGE gel of PCR products from DNA extracted from the biofilm – low chlorine dioxide phase of the chlorine dioxide experiment

Acquisition of Antibiotic Resistance in DWDS Biofilms

The data on antibiotic resistance acquisition is summarized in the following tables. As described above, it was measured using well characterized strains of *E. coli* that were allowed to establish in the biofilm and remain on the coupons during the experimental period. It should be noted that the antibiotic resistance was not measured in the biofilm but on suspensions taken in each case from the original isolates that had been frozen as a suspension; comparisons were made with the resistance of controls in all cases.

Some issues need discussion in presenting the results. First, the original screening in Halifax was performed on m-Endo agar, which is intended for total coliforms but which had been used previously in Halifax when working with *E. coli*. When the samples reached Ottawa, each colony showing a characteristic metallic sheen was transferred to culture in Mueller Hinton Agar and then its identity as *E. coli* was confirmed using mFC agar containing BCIG as substrate for the *E. coli* glucuronidase as described above. *E. coli* glucuronidase produces a blue colouration in the growing colony. Our results with m-FC containing BCIG medium confirm that in the untreated populations, the metallic sheen colonies from m-Endo were almost exclusively the blue *E. coli*. However, it is clear that in the treated samples the colonies growing on the m-FC with BCIG often contain a significant number, or even a majority, of white to off-white or pale yellow colonies – hereafter called 'whites'. Whereas it is ideal to screen very large numbers of colonies, the numbers of colonies in some cases were rather small with very few appearing in the higher disinfectant treatment level. Moreover, since the numbers of isolates obtained were not uniform, the numbers are not equivalent for each treatment condition.

Table 4.28 shows number of isolates from each reactor for each treatment, the total numbers of isolates examined as well as those that were glucuronidase positive, and also the percentage of total colonies isolated that were 'whites'. These numbers may differ slightly from the total numbers screened since a few of the colonies did not grow upon subculture. There are potentially two explanations for the appearance of white colonies after treatment. Either a significant number of the *E. coli* had been killed by the treatment and the other colonies showing up after treatment were organisms present originally in lower numbers but with a greater resistance to the applied disinfectant residual, or mutant *E. coli* had arisen that had damaged glucuronidase enzymes. The former was considered by far the more likely of the two explanations, and a few of the colonies were specifically identified, blinded as to whether they were blue or white, to confirm this. This identification data is shown below in Table 4.28

Table 4.28
The numbers of isolates screened and the percentage that were not *E. coli* from m-Endo plates

	Chlorine				Monochloramine				Chlorine Dioxide				
	untreated		treated		untreated		treated		untreated		treated		
	Total (B+W)	White as %	Total (B+W)	White as %	Total (B+W)	White as %	Total (B+W)	White as %	Total (B+W)	White as %	Total (B+W)	White %	Total numbers
AR1	384	0	205	1.0	1139	0	149	0.8	1142	0.0	1143	0.0	4162
AR2	384	0	384	0.3	1139	0	66	0.8	910	0.0	1104	0.0	3987
AR3	384	0	172	7.0	121	0	28	0.1	1143	0.1	136	0.0	1984
AR4	384	0	76	5.3	484	0	260	1.0	758	0.0	760	0.0	2722
AR5	384	0	382	1.0	1068	0	169	1.0	700	0.0	762	0.0	3465
AR6	383	0	382	0.5	1087	0	38	0.3	1085	0.0	762	0.5	3737
AR7	383	0	382	88.5	442	0	14	0.9	1105	0.2	635	30.2	2961
AR8	384	0	32	25.0	760	0	167	1.0	1143	0.0	648	0.0	3134
Total	3070		2015		6240		891		7986		5950		26152
Gluc+	3070		1645		6250		110		7983		6654		25712

An overall total of 25712 glucuronidase-positive isolates from treated and untreated reactors with chlorine, chloramine and chlorine dioxide were considered for analysis in this study. The isolates from each reactor/treatment were examined for their antibiotic resistance phenotypes against nine clinically relevant antibiotics following established protocols summarized in materials and methods section. The antibiotic selective pressures used for isolation of the resistant isolates were maintained at 2 to 4-fold above the known NCCLS MIC values for *E.coli* ATCC 25922 strain and our measured values determined for K12 and *mutS* strains in order to call the isolates resistant. The following three tables (Table 4.29 – 4.31) show the summary of results for the experiments with each of the disinfectants. In each of these tables the box on the left shows the number of isolates obtained without disinfectant treatment and the box on the right shows the isolates that were obtained following disinfectant treatment. For each disinfectant the table is in 3 parts: A - total numbers of multiple resistance isolates, B - the numbers of Rifampicin resistant isolates, and C - the Rifampicin resistant isolates as a percentage of the isolates obtained for each set of conditions. Certain trends are apparent. As expected, for all disinfectant experiments and conditions, the mismatch repair mutant strain usually produced much higher numbers of mutants than the wild type *E. coli*. For the chlorine experiment, the numbers of Rifampicin resistant isolates obtained for the *E. coli* K12 was generally very much less than for its isogenic mutant except for the polycarbonate substrate after chlorine treatment in the absence of upstream UV where the numbers for both wild type and mutant are similar; no isolates were obtained with the wild-type *E. coli* that were resistant to more than 3 antibiotics, but the numbers were low. A somewhat similar pattern was seen for chloramines except that more multiply-resistant isolates were obtained even for wild type treated isolates. Chlorine dioxide also produced higher numbers of wild-type mutants than chlorine, especially in AR7 (polycarbonate in the absence of upstream UV. It is not known whether these apparent aberrations for polycarbonate substrate and no UV treatment are real or are an unknown artifact of that annular reactor.

Using our established databases the antibiotic resistant isolates were categorized into phenotypic groups each with unique multiple resistance characteristics. From these categories, representative members were considered for further molecular analyses. Regardless of the treatment and the type of the biofilm substratum used, the overall effect of the *mutS* deficiency, i.e. MMR dysfunction, on co-occurrence of Rif^r marker among multiple resistant (up to 9 antibiotics) isolates was 3- to 14-fold higher than among K12 isolates (Table 4.32). This signifies the role of the mismatch repair deficiency in contributing to emergence of multiple antibiotic resistances among *E. coli* isolates.

Table 4.29
Multiple antibiotics resistant *E. coli K12* and *E. coli.* muts isolates from untreated control and chlorine treated biofilm reactors

Chlorine treatment and occurrence of multiple antibiotic resistance among *E.coli K12* and *E.coli mutS* strains

A)

Multiple antibiotic resistant isolates (untreated)

#of antibiotics	AR1	AR2	AR3	AR4	AR5	AR6	AR7	AR8
1	7	176	6	169	0	163	4	147
2	0	104	0	25	0	29	7	29
3	0	6	0	2	0	0	1	2
4	0	0	0	0	0	0	0	0
5	0	0	0	0	0	0	0	0
6	0	0	0	0	0	0	0	0
7	0	0	0	0	0	0	0	0
8	0	0	0	0	0	0	0	0
9	0	0	0	0	0	0	0	0
Total	7	286	6	196	0	192	12	178
Chlorine	-	-	-	-	-	-	-	-
Cast iron	+	+	-	-	+	+	-	-
Polycarbonate	-	-	+	+	-	-	+	+
UV treated water	+	+	+	+	-	-	-	-
WT	+	-	+	-	+	-	+	-
mutS	-	+	-	+	-	+	-	+

Chlorine treated multiple resistant isolates

#of antibiotics	AR1	AR2	AR3	AR4	AR5	AR6	AR7	AR8
1	5	181	7	20	4	277	15	11
2	0	76	1	26	0	78	4	7
3	0	3	0	18	0	6	1	3
4	0	0	0	4	1	0	0	0
5	0	0	0	2	0	0	0	1
6	0	0	0	1	0	0	0	0
7	0	0	0	0	0	0	0	0
8	0	0	0	0	0	0	0	0
9	0	0	0	0	0	0	0	0
Total	5	260	8	71	5	361	20	22
Chlorine	+	+	+	+	+	+	+	+
Cast iron	+	+	-	-	+	+	-	-
Polycarbonate	-	-	+	+	-	-	+	+
UV treated water	+	+	+	+	-	-	-	-
WT	+	-	+	-	+	-	+	-
mutS	-	+	-	+	-	+	-	+

B)

Rifampicin resistant Isolates (untreated)

#of antibiotics	AR1	AR2	AR3	AR4	AR5	AR6	AR7	AR8
1	1	57	4	115	0	138	4	95
2	0	96	0	25	0	28	5	23
3	0	6	0	2	0	0	0	2
4	0	0	0	0	0	0	0	0
5	0	0	0	0	0	0	0	0
6	0	0	0	0	0	0	0	0
7	0	0	0	0	0	0	0	0
8	0	0	0	0	0	0	0	0
9	0	0	0	0	0	0	0	0
Total	1	159	4	142	0	166	9	120
Chlorine	-	-	-	-	-	-	-	-
Cast iron	+	+	-	-	+	+	-	-
Polycarbonate	-	-	+	+	-	-	+	+
UV treated water	+	+	+	+	-	-	-	-
WT	+	-	+	-	+	-	+	-
mutS	-	+	-	+	-	+	-	+

Rifampicin resistant Isolates (treated)

#of antibiotics	AR1	AR2	AR3	AR4	AR5	AR6	AR7	AR8
1	2	130	1	13	1	269	1	8
2	0	51	0	24	0	78	1	7
3	0	3	0	17	0	6	0	3
4	0	0	0	4	0	0	0	0
5	0	0	0	0	0	0	0	0
6	0	0	0	1	0	0	0	0
7	0	0	0	0	0	0	0	0
8	0	0	0	0	0	0	0	0
9	0	0	0	0	0	0	0	0
Total	2	184	1	59	1	353	2	18
Chlorine	+	+	+	+	+	+	+	+
Cast iron	+	+	-	-	+	+	-	-
Polycarbonate	-	-	+	+	-	-	+	+
UV treated water	+	+	+	+	-	-	-	-
WT	+	-	+	-	+	-	+	-
mutS	-	+	-	+	-	+	-	+

C)

Percentage of rifampicin resistant isolates in untreated reactors

#of antibiotics	AR1	AR2	AR3	AR4	AR5	AR6	AR7	AR8
1	14	32	67	68	NA	85	100	65
2	NA	92	NA	100	NA	97	71	79
3	NA	100	NA	100	NA	NA	0	100
4	NA	NA	NA	NA	NA	NA	NA	NA
5	NA	NA	NA	NA	NA	NA	NA	NA
6	NA	NA	NA	NA	NA	NA	NA	NA
7	NA	NA	NA	NA	NA	NA	NA	NA
8	NA	NA	NA	NA	NA	NA	NA	NA
9	NA	NA	NA	NA	NA	NA	NA	NA
Chlorine	-	-	-	-	-	-	-	-
Cast iron	+	+	-	-	+	+	-	-
Polycarbonate	-	-	+	+	-	-	+	+
UV treated water	+	+	+	+	-	-	-	-
WT	+	-	+	-	+	-	+	-
mutS	-	+	-	+	-	+	-	+

NA: *not applicable due to lack of isolate*

Percentage of rifampicin resistant isolates in treated reactors

#of antibiotics	AR1	AR2	AR3	AR4	AR5	AR6	AR7	AR8
1	40	72	14	65	25	97	7	73
2	NA	67	0	92	NA	100	25	100
3	NA	100	NA	94	NA	100	0	100
4	NA	NA	NA	100	0	NA	NA	NA
5	NA	NA	NA	0	NA	NA	NA	0
6	NA	NA	NA	100	NA	NA	NA	NA
7	NA	NA	NA	NA	NA	NA	NA	NA
8	NA	NA	NA	NA	NA	NA	NA	NA
9	NA	NA	NA	NA	NA	NA	NA	NA
Chlorine	+	+	+	+	+	+	+	+
Cast iron	+	+	-	-	+	+	-	-
Polycarbonate	-	-	+	+	-	-	+	+
UV treated water	+	+	+	+	-	-	-	-
WT	+	-	+	-	+	-	+	-
mutS	-	+	-	+	-	+	-	+

Table 4.30

Multiple antibiotics resistant *E. coli K12* and *E. coli.* muts isolates from untreated control and chloramine treated biofilm reactors

Chloramine treatment and occurrence of multiple antibiotic resistance among *E.coli K12* and *E.coli mutS* strains

A)

Multiple antibiotic resistant isolates (untreated)

#of antibiotics	AR1	AR2	AR3	AR4	AR5	AR6	AR7	AR8
1	201	357	92	114	767	381	82	240
2	7	443	17	234	31	503	29	290
3	0	139	5	115	4	154	18	95
4	0	5	1	6	2	9	11	16
5	0	0	1	0	0	2	11	2
6	0	0	3	3	0	1	9	4
7	0	0	1	1	0	0	3	1
8	0	0	2	1	0	0	0	1
9	0	0	0	2	0	0	0	0
Total	208	944	122	476	804	1050	163	649
Chloramine	-	-	-	-	-	-	-	-
Cast iron	+	+	-	-	+	+	-	-
Polycarbonate	-	-	+	+	-	-	+	+
UV treated water	+	+	+	+	-	-	-	-
WT	+	-	+	-	+	-	+	-
mutS	-	+	-	+	-	+	-	+

Chloramine treated multiple resistant isolates

#of antibiotics	AR1	AR2	AR3	AR4	AR5	AR6	AR7	AR8
1	5	1	7	0	0	1	2	0
2	6	0	2	0	0	10	0	0
3	5	1	0	1	0	6	0	0
4	2	0	1	0	0	5	0	0
5	12	0	2	0	0	2	0	1
6	7	0	0	0	0	3	0	1
7	1	2	0	0	0	0	0	0
8	0	6	1	0	0	0	0	0
9	0	3	1	0	0	0	0	0
Total	38	13	14	1	0	27	2	2
Chloramine	+	+	+	+	+	+	+	+
Cast iron	+	+	-	-	+	+	-	-
Polycarbonate	-	-	+	+	-	-	+	+
UV treated water	+	+	+	+	-	-	-	-
WT	+	-	+	-	+	-	+	-
mutS	-	+	-	+	-	+	-	+

B)

Rifampicin resistant Isolates (untreated)

#of antibiotics	AR1	AR2	AR3	AR4	AR5	AR6	AR7	AR8
1	8	205	0	36	4	137	3	205
2	3	281	1	170	20	257	2	143
3	0	63	0	90	0	105	1	25
4	0	3	0	2	0	4	0	6
5	0	0	1	2	0	0	1	1
6	0	0	0	1	0	0	2	3
7	0	0	0	1	0	0	0	1
8	0	0	0	1	0	0	0	1
9	0	0	0	2	0	0	0	0
Total	11	552	2	305	24	503	9	385
Chloramine	-	-	-	-	-	-	-	-
Cast iron	+	+	-	-	+	+	-	-
Polycarbonate	-	-	+	+	-	-	+	+
UV treated water	+	+	+	+	-	-	-	-
WT	+	-	+	-	+	-	+	-
mutS	-	+	-	+	-	+	-	+

Rifampicin resistant Isolates (treated)

#of antibiotics	AR1	AR2	AR3	AR4	AR5	AR6	AR7	AR8
1	0	0	0	0	0	0	0	0
2	1	0	0	0	0	7	0	0
3	0	0	0	0	0	0	0	0
4	2	0	0	0	0	2	0	0
5	3	0	0	0	0	1	0	0
6	2	0	0	0	0	3	0	1
7	1	2	0	0	0	0	0	0
8	0	6	1	0	0	0	0	0
9	0	3	1	0	0	0	0	0
Total	9	11	2	0	0	13	0	1
Chloramine	+	+	+	+	+	+	+	+
Cast iron	+	+	-	-	+	+	-	-
Polycarbonate	-	-	+	+	-	-	+	+
UV treated water	+	+	+	+	-	-	-	-
WT	+	-	+	-	+	-	+	-
mutS	-	+	-	+	-	+	-	+

C)

Percentage of rifampicin resistant isolates in untreated reactors

#of antibiotics	AR1	AR2	AR3	AR4	AR5	AR6	AR7	AR8
1	4	57	0	32	1	36	4	85
2	43	63	6	73	65	51	7	49
3	NA	45	0	78	0	68	6	26
4	NA	60	0	33	0	44	0	38
5	NA	NA	100	NA	NA	0	9	50
6	NA	NA	0	33	NA	0	22	75
7	NA	NA	0	100	NA	NA	0	100
8	NA	NA	0	100	NA	NA	NA	100
9	NA	NA	NA	100	NA	NA	NA	NA
Chloramine	-	-	-	-	-	-	-	-
Cast iron	+	+	-	-	+	+	-	-
Polycarbonate	-	-	+	+	-	-	+	+
UV treated water	+	+	+	+	-	-	-	-
WT	+	-	+	-	+	-	+	-
mutS	-	+	-	+	-	+	-	+

Percentage of rifampicin resistant isolates in treated reactors

#of antibiotics	AR1	AR2	AR3	AR4	AR5	AR6	AR7	AR8
1	0	0	0	NA	NA	0	0	NA
2	17	NA	0	NA	NA	70	NA	NA
3	0	0	NA	0	NA	0	NA	NA
4	100	NA	0	NA	NA	40	NA	NA
5	25	NA	0	NA	NA	50	NA	0
6	29	NA	NA	NA	NA	100	NA	100
7	100	100	NA	NA	NA	NA	NA	NA
8	NA	100	100	NA	NA	NA	NA	NA
9	NA	100	100	NA	NA	NA	NA	NA
Chloramine	+	+	+	+	+	+	+	+
Cast iron	+	+	-	-	+	+	-	-
Polycarbonate	-	-	+	+	-	-	+	+
UV treated water	+	+	+	+	-	-	-	-
WT	+	-	+	-	+	-	+	-
mutS	-	+	-	+	-	+	-	+

NA: *not applicable due to lack of isolate*

Table 4.31
Multiple antibiotics resistant *E. coli K12* and *E. coli.* muts isolates from untreated control and chlorine dioxide treated biofilm reactors

Chlorine dioxide treatment and occurrence of multiple antibiotic resistance among *E.coli K12* and *E.coli mutS* strains

A)

Multiple antibiotic resistant isolates (untreated)

#of antibiotics	AR1	AR2	AR3	AR4	AR5	AR6	AR7	AR8
1	29	265	55	130	11	235	42	101
2	0	79	1	22	0	40	10	81
3	0	0	0	0	0	0	2	70
4	0	0	0	3	0	0	3	39
5	0	0	0	5	0	0	2	35
6	0	0	0	13	0	0	5	66
7	0	0	0	0	0	0	6	7
8	0	0	0	0	0	0	0	0
9	0	0	0	0	0	0	0	0
Total	29	344	56	173	11	275	70	399
Chlorine dioxide	-	-	-	-	-	-	-	-
Cast iron	+	+	-	-	+	+	-	-
Polycarbonate	-	-	+	+	-	-	+	+
UV treated water	+	+	+	+	-	-	-	-
WT	+	-	+	-	+	-	+	-
mutS	-	+	-	+	-	+	-	+

Chlorine dioxide treated multiple resistant isolates

#of antibiotics	AR1	AR2	AR3	AR4	AR5	AR6	AR7	AR8
1	13	106	53	66	9	122	59	161
2	0	6	3	17	0	147	75	134
3	0	3	0	5	0	48	27	192
4	0	1	0	0	0	24	37	46
5	0	1	0	0	0	17	106	49
6	0	1	0	0	0	21	79	11
7	0	0	0	0	0	19	31	1
8	0	0	0	0	0	57	1	0
9	0	0	0	0	0	0	0	0
Total	13	118	56	88	9	455	415	594
Chlorine dioxide	+	+	+	+	+	+	+	+
Cast iron	+	+	-	-	+	+	-	-
Polycarbonate	-	-	+	+	-	-	+	+
UV treated water	+	+	+	+	-	-	-	-
WT	+	-	+	-	+	-	+	-
mutS	-	+	-	+	-	+	-	+

B)

Rifampicin resistant Isolates (untreated)

#of antibiotics	AR1	AR2	AR3	AR4	AR5	AR6	AR7	AR8
1	14	210	15	86	10	204	12	51
2	0	79	1	19	0	39	1	25
3	0	0	0	0	0	0	0	9
4	0	0	0	0	0	0	0	3
5	0	0	0	1	0	0	0	2
6	0	0	0	0	0	0	0	4
7	0	0	0	0	0	0	0	5
8	0	0	0	0	0	0	0	0
9	0	0	0	0	0	0	0	0
Total	14	289	16	106	10	243	13	99
Chlorine dioxide	-	-	-	-	-	-	-	-
Cast iron	+	+	-	-	+	+	-	-
Polycarbonate	-	-	+	+	-	-	+	+
UV treated water	+	+	+	+	-	-	-	-
WT	+	-	+	-	+	-	+	-
mutS	-	+	-	+	-	+	-	+

Rifampicin resistant Isolates (treated)

#of antibiotics	AR1	AR2	AR3	AR4	AR5	AR6	AR7	AR8
1	9	23	1	34	8	18	1	0
2	0	4	1	15	0	3	0	15
3	0	0	0	4	0	1	0	4
4	0	0	0	0	0	0	0	5
5	0	0	0	0	0	2	0	4
6	0	0	0	0	0	1	4	4
7	0	0	0	0	0	1	15	0
8	0	0	0	0	0	3	1	0
9	0	0	0	0	0	0	0	0
Total	9	27	2	53	8	29	21	32
Chlorine dioxide	+	+	+	+	+	+	+	+
Cast iron	+	+	-	-	+	+	-	-
Polycarbonate	-	-	+	+	-	-	+	+
UV treated water	+	+	+	+	-	-	-	-
WT	+	-	+	-	+	-	+	-
mutS	-	+	-	+	-	+	-	+

C)

Percentage of rifampicin resistant isolates in untreated reactors

#of antibiotics	AR1	AR2	AR3	AR4	AR5	AR6	AR7	AR8
1	48	79	27	66	91	87	29	50
2	NA	100	100	86	NA	98	10	31
3	NA	NA	NA	NA	NA	NA	0	13
4	NA	NA	NA	0	NA	NA	0	8
5	NA	NA	NA	20	NA	NA	0	6
6	NA	NA	NA	0	NA	NA	0	6
7	NA	NA	NA	NA	NA	NA	0	71
8	NA	NA	NA	NA	NA	NA	NA	NA
9	NA	NA	NA	NA	NA	NA	NA	NA
Chlorine dioxide	-	-	-	-	-	-	-	-
Cast iron	+	+	-	-	+	+	-	-
Polycarbonate	-	-	+	+	-	-	+	+
UV treated water	+	+	+	+	-	-	-	-
WT	+	-	+	-	+	-	+	-
mutS	-	+	-	+	-	+	-	+

NA: *not applicable due to lack of isolate*

Percentage of rifampicin resistant isolates in treated reactors

#of antibiotics	AR1	AR2	AR3	AR4	AR5	AR6	AR7	AR8
1	69	22	2	52	89	15	2	0
2	NA	67	33	88	NA	2	0	11
3	NA	0	NA	80	NA	2	0	2
4	NA	0	NA	NA	NA	0	0	11
5	NA	0	NA	NA	NA	12	0	8
6	NA	0	NA	NA	NA	5	5	36
7	NA	NA	NA	NA	NA	5	48	0
8	NA	NA	NA	NA	NA	5	100	NA
9	NA	NA	NA	NA	NA	NA	NA	NA
Chlorine dioxide	+	+	+	+	+	+	+	+
Cast iron	+	+	-	-	+	+	-	-
Polycarbonate	-	-	+	+	-	-	+	+
UV treated water	+	+	+	+	-	-	-	-
WT	+	-	+	-	+	-	+	-
mutS	-	+	-	+	-	+	-	+

Table 4.32
Overall relative co-occurrence of Rifampicin-resistant phenotype among multiple resistant isolates

A)

Chloramine	Untreated		Treated			
Strain	K12	mutS	K12	mutS	Overall K12	overall mutS
Resistant isolates	1297	3119	54	43	1351	3162
Rifampicin resistant	46	1745	11	25	57	1770
%Rif^r of total	4	56	20	58	4	56

B)

Chlorine	Untreated		Treated			
Strain	K12	mutS	K12	mutS	Overall K12	overall muts
Resistant isolates	25	852	38	714	63	1566
Rifampicin resistant	14	587	6	614	20	1201
%Rif^r of total	56	69	16	86	32	77

C)

Chlorine dioxide	Untreated		Treated			
Strain	K12	mutS	K12	mutS	Overall K12	overall muts
Resistant isolates	166	1191	493	1255	659	2446
Rifampicin resistant	53	737	40	141	93	878
%Rif^r of total	32	62	8	11	14	36

A total of 1297 and 3119 isolates originated from untreated wild type K12 and *mutS* strains, respectively, together with 54 and 43 isolates from chloramine treated wild type K12 and mutS origin, respectively, demonstrated resistance against 1-9 antibiotics (Table 4.33A) Overall, 56% of the 1-9 antibiotic resistant *mutS* isolates among untreated cells were also resistant to rifampicin. This value was slightly increased (58%) for the chloramine treated reactors. In agreement with previous studies, these high values are indicative of a direct correlation between mismatch repair deficiency and the occurrence of the Rif^r phenotype. Since *mutS* strain is devoid of mismatch repair functions, occurrence of mutagenic events at *rpoB* gene, which is a genetic target exclusively specific for base substitution mutagenesis, would already be saturated due to mismatch deficiency. Therefore, it is reasonable to consider that chloramine treatment would have not been able to generate significant increase in mutational recovery at this target among isolates of *mutS* origin. The behaviour of the *mutS* strain in untreated, chlorine and chlorine dioxide treated reactors follows almost a similar pattern. In untreated, chlorine treated and the control set for chlorine dioxide treatment, 62-86% of multiple resistant isolates were also resistant to rifampicin. However, it is not clear why this value for the chlorine dioxide treated cells was low (Table 4-33 B and C). In contrast, the mismatch repair proficient K12 strain generated rifampicin co-resistant isolates at the rates of 4% and 20% of the total of 1-9 antibiotic-resistant populations in untreated and chloramines treated cells, respectively. This accounts for a 5-fold increase in rifampicin resistant isolates among chloramine-treated populations of which a major subpopulation is expected to be mismatch repair deficient. However, although no similar effect was observed among chlorine and chlorine dioxide treated biofilms, this does not exclude the possibility of occurrence of *mmr* phenotype among the present isolates from K12 origin, which need to be characterized further. This is discussed further below.

Table 4.33
Use of cast iron (CI) or polycarbonate (PC) coupons in reactors and the coincidence of resistance to Rifampicin and multiple antibiotics for reactors with upstream UV treatment

A)

Chloramine	Untreated K12		Treated K12		Untreated mutS		Treated mutS		Overall K12		Overall mutS	
Coupon	CI	PC	CI	PC	CI	PC	CI	PC	CI	PC	CI	PC
Resistant isolates	208	122	38	14	944	476	13	1	246	136	957	477
Rifampicin resistant	11	2	9	2	552	305	11	0	20	4	563	305
%Rif^r of total	5	2	24	14	58	64	85	0	8	3	59	64

B)

Chlorine	Untreated K12		Treated K12		Untreated mutS		Treated mutS		Overal K12		Overall mutS	
Coupon	CI	PC	CI	PC	CI	PC	CI	PC	CI	PC	CI	PC
Resistant isolates	7	6	5	8	286	196	260	71	12	14	546	267
Rifampicin resistant	1	4	2	1	159	142	184	59	3	5	343	201
%Rif^r of total	14	67	40	13	56	72	71	83	25	36	63	75

C)

Chlorine dioxide	Untreated K12		Treated K12		Untreated mutS		Treated mutS		Overall K12		Overall mutS	
Coupon	CI	PC	CI	PC	CI	PC	CI	PC	CI	PC	CI	PC
Resistant isolates	29	56	13	56	344	173	118	88	42	112	462	261
Rifampicin resistant	14	16	9	2	289	106	27	53	23	18	316	159
%Rif^r of total	48	29	69	4	84	61	23	60	55	16	68	61

In order to see whether a subpopulation of *mmr* mutants play a role in the emergence and evolution of the multiple antibiotic resistance in bacterial biofilm, we categorized the resistant isolates on the basis of their resistance patterns against 1 to 9 antibiotics and determined the number of the isolates that were also resistant to rifampicin in each category. These results are organized in Tables 4.29, 4.30 and 4.31. In each table, among the treated and untreated sets, the total number of the resistant isolates against 1-9 antibiotics (shown in A) and the number of the rifampicin resistant isolates corresponding to each category (shown in B) and the relative proportion of the isolates within each category that were also resistant against rifampicin are summarized in C. Regardless of the type of treatment and the biofilm coupons used, a positive correlation between the relative co-occurrence of rifampicin resistant mutants from *mutS* origin (even numbered reactors) and the multiplicity of antibiotic resistance among resistant isolates was apparent. In many cases the relative number of the multiply resistant isolates that were also resistant to rifampicin approached the total number of the isolates in that category as they acquired resistance towards higher number of antibiotics. This would mean that in the absence of mismatch repair functions the likelihood of rifampicin resistant mutants, the major *mmr* phenotype marker, to develop resistance against other antibiotics while the cells were growing in biofilm is *mmr* phenotype dependent (Tables 4.29C, 4.30C and 4.31C). An additional shift towards higher antibiotic resistance multiplicity is apparent among the treated cells as compared to untreated reactors with the most pronounced shift among chloramine treated cells. More or less, a similar pattern was observed among the isolates from the K12 origin, which is an indication of a possible role of an *mmr* mutant subpopulation in developing multiple resistance phenotype among these isolates (Tables 4.29C, 4.30C and 4.31C).

Neither upstream UV treatment of the water supplied into the reactors nor the nature of the coupon material that were used in reactors had any obvious and consistent effect on the relative numbers of isolates with rifampicin resistant phenotype among the multiply resistant isolates (Table 4.33 and Table 4.34). This would mean that neither of these factors had any clear contribution to mutational events that might have been resulted in development of the candidate *mmr* mutants and further development of multiple resistances among isolates. Though beyond the agreed scope of this study, in order to find out what proportion of the multiply resistant isolates that carry rifampicin resistance mutation are also mismatch repair deficient, at least two complementary approaches are being taken.

Considering that the spontaneous mutational spectra of the wild type K12 and mismatch repair deficient *mutS* strains at *rpoB* gene target are known, one approach would be to clone the *rpoB* gene from the untreated and treated isolates that are Rifampicin resistant. Sequencing these clones and comparing them to the mutational spectra of wild type and *mutS* strains would reveal useful information about the involvement of the mismatch repair deficiency. It is also expected that analysis of the sequence data will reveal additional information regarding the nature of the treatment-induced mutations at *rpoB* gene. We have optimized the PCR conditions to amplify the cluster II region of the *rpoB* gene using the specific primers designed to be used on genomic templates as shown in Figure 3.7. Under the optimal conditions a single band of 636 bp was produced using the genomic templates from control wild type K12 and *mutS* strains (Figure 4.41A). This PCR product was further verified by *BclI* digestion that is expected to produce two smaller bands of 227 and 409 bp. Interestingly, during this verification we noticed that the PCR product generated from a colony from *mutS* strain lacked the digestion site for *BclI* and that this site is one of the 69 hotspots that is frequently mutated in *rpoB* gene (Figure 4.41B). Again this accidental observation confirms the direct relationship between *mmr* strains and high correlation with base substitution events in *rpoB* gene.

The second approach will consider western blotting of cell lysates from candidate *mmr* isolates using anti MutS, MutL and MutH antibodies. This approach will identify the *mmr* mutants that occur due to truncation of any of MutS, MutL and MutH proteins. Using the sizes of the truncated proteins it is possible to estimate the region of the genes where the mutation has occurred. Specific primers flanking the region will be used to sequence the region to identify the nature of the mutation. Inactivating mutations in mutSHL genes among the remaining *mmr* candidates that do not show any truncation of MutSLH proteins will be identified by cloning and direct sequencing of the relevent genes.

One final issue needs mentioning in the context of studying the acquisition of antibiotic resistance, and that is the relatively large number of mutations seen in the untreated samples. Certainly there are multiple influences at play here. Firstly there are certainly a lot more cells present initially and therefore a much higher cell population to select from. Also, since the cells are likely to be in stationary phase due to starvation pressure, adaptive mutation is much more prevalent under these conditions. Finally, there is an unknown influence that may remain from residual DBP's that could be in the system. The source water for these experiments was dechlorinated tap water from Halifax and the efficiency of the activated carbon columns at removing the DBP from the incoming water is unknown. Which of these is the most influential is unknown and indeed all could play a role in the numbers of resistant isolates observed.

Table 4.34
Relative occurrence of rif co-resistant isolates among multiple resistant isolates

	Overall K12		Overall mutS		Overall K12		**Overall mutS**	
Coupon type	CI	PC	CI	PC	CI	PC	CI	PC
Water UV treatment	+	+	+	+	-	-	-	-
Chloramine	8	3	59	64	3	5	48	59
Chlorine	25	36	63	75	20	34	94	69
chlorine dioxide	55	16	68	61	90	7	37	13

*Summary of coupon type and UV treatment of water:

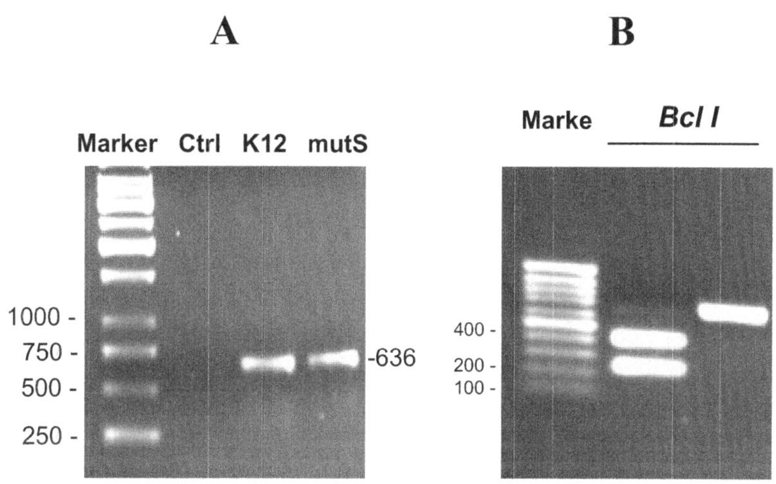

Figure 4.41 PCR products generated from cluster II of rpoB gene (A); BclI digestion of PCR products (B)

SUMMARY

This chapter examined the effect of UV treatment, secondary disinfection and pipe material on two effluent and biofilm *E. coli* strains: *E. coli K12* and a derived mutant strain. In addition, phenotypical responses were examined under these conditions to see if the *E. coli* adapted or changed phenotypically under certain environmental conditions.

In general biofilm *E. coli* was more resistant to chemical disinfection than *E. coli* in suspension. Several studies confirm these findings as biofilms tend to protect organisms from disinfection while planktonic bacteria in the bulk fluid tend to perish more quickly (Camper and Jones 2000, Skinner 2003, and Parsek and Fuqua 2004).

The *E. coli K12* strain was much more resistant to UV disinfection (Post-UV) than its mutant counterpart. Mutant colonies were never recovered from the effluent or biofilm of either Post-UV treated cast iron or polycarbonate annular reactors. In terms of effluent *E. coli,* there were no apparent differences between the two strains. In contrast, biofilm *E. coli K12* seemed to be more resistant to disinfection than the mutant strain in chlorinated polycarbonate reactors and cast iron chloraminated reactors. There were no variations between the two species in ARs that were treated with chlorine dioxide.

UV disinfection (Post-UV) was extremely effective against *E. coli* that was spiked through the UV unit. Comparable effects were seen on effluent and biofilm colonies on both polycarbonate and cast iron surfaces. Pre-UV treated water had no obvious benefit over the Non-UV treated water in terms of added protection against a breech in distribution integrity. In the simulated intrusion (spike) there were no evident trends indicating that *E. coli* could attach itself better or worse in a UV treated aquatic environment. These findings are consistent with those of other studies (Pozos et al. 2004).

Pipe material had little effect on effluent *E. coli* in both the acclimation and disinfection phases of the experiments. During the acclimation phase there were also few significant differences among biofilm *E. coli* on both polycarbonate and cast iron coupons. However, during disinfection, there were several cases where *E. coli* on cast iron surfaces was more resistant to disinfection than those on polycarbonate surfaces which are consistent with the findings of others (LeChevallier et al.1996, and Camper et al. 1999). In addition, cast iron reactors tended to have increased turbidity as a result of iron sloughing off the coupon surfaces. Consequently, disinfectant demands rose and often made it difficult to maintain disinfectant residuals, therefore making it easier for coliforms to persist within cast iron biofilms.

There were nominal variations among secondary disinfectants regarding effluent *E. coli* reduction in model distribution systems. Significant distinctions were noted during chemical disinfection of Pre and Non-UV cast iron reactor biofilms. Overall, chlorine followed by monochloramine and then chlorine dioxide were the most to least effective, respectively, at decreasing biofilm *E. coli* on cast iron surfaces. *E. coli* on polycarbonate surfaces were relatively equally decreased with all three disinfectants. For Post-UV reactors, log reductions in effluent and biofilm *E. coli K12* were statistically comparable between chlorine, chlorine dioxide and chloramines.

Overall, phenotypic responses of *E. coli* isolated from the AR biofilms varied based on pipe material and chemical disinfectant. It was concluded that there are several explanations to why *E. coli* isolated from the annular reactor biofilms differed in phenotypical profile (API 20E test) from the original stock culture used to simulate a breech in distribution system integrity. Some factors that may have influenced these phenotypical changes include: diversification of organisms within the biofilm communities and sublethal injury as a result of chemical disinfection. In addition, UV treatment and pipe material may have influenced them to change phenotypically in order to survive and cope in a different environment. As a result, they may have developed a "biofilm phenotype" which varied from that of the stock and planktonic culture (Parsek and Fuqua 2004).

The biofilm composition studies are certainly not definitive of what would happen in any system but a snapshot of what is happening in the experimental system studied and they reflect the composition of the incoming water as well as the conditions applied. With a relatively constant water composition one would certainly expect some similarities to occur over time. To some extent this is seen but there are also different disinfectants used that can affect the composition of the biofilm community. Nevertheless, seasonal effects can occur in species succession and it is possible that some of the changes seen in Pseudomonas species could have been seasonal effects. That multiple isolations seem to occur suggests a certain level of species dominance that is characteristic of many molecular detection methods. Nonetheless, there are clearly potentials for biases in the molecular methods that are unavoidable and must be accepted as potential interference with the true picture. Comparisons between substrates show some clear differences of organisms that can grow on one substrate but not on the other. Similarly for disinfectant treatments, including UV, differences can be seen and in some cases organisms are identified at higher doses that were not seen at lower ones. This is presumably due to intrinsic resistance but combined with an inability to complete numerically with some of the more sensitive organisms. Since this is one of the first studies of its type in drinking water, it will be interesting to observe what is seen in future studies and make comparisons.

Antibiotic resistance can arise from many factors. The likely candidates here are mismatch mutations occurring during conditions of starvation and redox stress, or genotoxicity from disinfectants, disinfection by-products or other oxidized or toxic moieties that can cumulatively result in mutations. Once again this is a very early study of its type and the mechanistic and exact causation has not been sufficiently explored to give definitive answers. Clearly more studies are desirable and should be conducted under more controlled conditions, but the fact the antibiotic resistant mutants, that were commonly resistant to a multiplicity of antibiotics, occurred for all treatment conditions, including controls shows an important proof of concept and needs further investigation. Nevertheless the current data suggest that progenitors for antibiotic resistant phenotypes can be selected for by continuous selective pressure s in the disitribution system. For this part of the study there was no apparent effect of pipe material or upstream UV.

CHAPTER 5
FIELD SCALE DISINFECTION STUDIES – SURFACE WATER

INTRODUCTION

The Halifax Regional Water Commission (HRWC) services over 310,000 people in the greater Halifax area in Nova Scotia. The J.D Kline Water Supply Plant is one of two major plants operated by HRWC that has a capacity of 220 ML/day and treats raw surface water from Pockwock Lake in Halifax, Nova Scotia. The treatment process consists of direct filtration with chlorination for final disinfection. The source water is low in alkalinity (<10 mg/L as $CaCO_3$) and typically has no occurrences of *Cryptosporidium parvum* or *Giardia lamblia* since HRWC owns the land surrounding their watershed. Although no significant problems plague the operation of the plant in its current state, new Nova Scotia regulations will be implemented by 2008 and the HRWC wants to ensure that practices result in absolute compliance at that time. With chlorination of surface water, the potential for TTHM and HAA formation is a concern. It is possible that changing to UV and/or ClO_2 disinfection may reduce these DBP concentrations to a minimal level. In addition, HRWC was interested in the effectiveness of different disinfection strategies against heterotrophic bacteria including *Escherichia coli* in the distribution system. The focus of this chapter was to determine the effectiveness of UV, UV/ Cl_2, and UV/ ClO_2 in suppressing biofilm growth and controlling microbial pathogens in drinking water distribution systems through a field-scale study, and to compare pre-treatment with UV to no UV treatment. Heterotrophic plate counts and coliforms were used to compare disinfection treatments for bacteria within the systems. In addition, water quality parameters were monitored throughout the experiments.

RESULTS AND DISCUSSION

Acclimation/Pre-disinfection period

After a three-week acclimation period one AR that was receiving UV-treated water started being disinfected with chlorine and one AR pre-treated with UV was disinfected with chlorine dioxide. Two ARs that had no UV disinfection were also disinfected with either Cl_2, or ClO_2. The remaining two ARs were not altered and continued as experimental controls, one for UV water and one for raw water. Following the first trial run, the experiment was repeated under identical conditions, allowing for a re-acclimation period between runs. Data presented represent the combined acclimation and disinfection stages of both trial runs. Before disinfection the overall mean number of suspended heterotrophic bacteria was $3.48 \times 10^4 \pm 3.19 \times 10^4$ CFU/mL and mean attached (biofilm) heterotrophic bacteria for all ARs was $6.20 \times 10^4 \pm 4.70 \times 10^4$ CFU/cm^2. No ARs were statistically different for heterotrophic bacteria in the acclimation period for either suspended or attached cells.

Post-disinfection HPC data

All ARs that were treated with Cl_2 or ClO_2 showed a decrease in suspended heterotrophic bacteria, especially those treated with ClO_2 (Figure 5.1). There was not a significant difference between average suspended HPC bacteria pre- and post-disinfection, however each AR showed reduction with an average count of 1.08x104± 2.05x104 CFU/mL in all ARs after disinfection, which decreased from 3.48x104± 3.19x104 CFU/mL (Table 5.1). The average log reduction for suspended HPC bacteria for ARs treated with chemical disinfection was 1.33 ± 0.73. ARs treated with ClO_2 had the greatest log reductions including 2.12 for the AR treated with both UV and chlorine dioxide, and 1.64 for the AR treated with ClO_2 only. The AR treated with UV and chlorine had a log reduction of 1.15 for suspended HPC bacteria and similarly the AR treated with Cl_2 had an average log reduction of 0.40. Overall the ARs treated with UV had greater reductions in suspended bacteria than those ARs treated with corresponding chemical disinfection only.

Table 5.1
Suspended and Attached HPC Bacteria Pre- and Post-Disinfection

AR	Suspended HPC (CFU/mL) Acclimation	Attached HPC (CFU/cm^2) Acclimation	Suspended HPC (CFU/mL) Post-Disinfection	Attached HPC (CFU/cm^2) Post-Disinfection
UV Control	9.44×10^4	9.86×10^4	5.24×10^4	1.89×10^5
RW Control	2.42×10^4	3.20×10^4	5.57×10^3	5.84×10^4
UV+ClO_2	3.46×10^4	8.86×10^4	2.63×10^2	3.26×10^3
ClO_2	7.32×10^3	8.87×10^3	1.66×10^2	1.87×10^3
UV+Cl_2	3.90×10^4	1.22×10^5	2.79×10^3	2.43×10^3
Cl_2	9.38×10^3	2.15×10^4	3.71×10^3	3.90×10^3

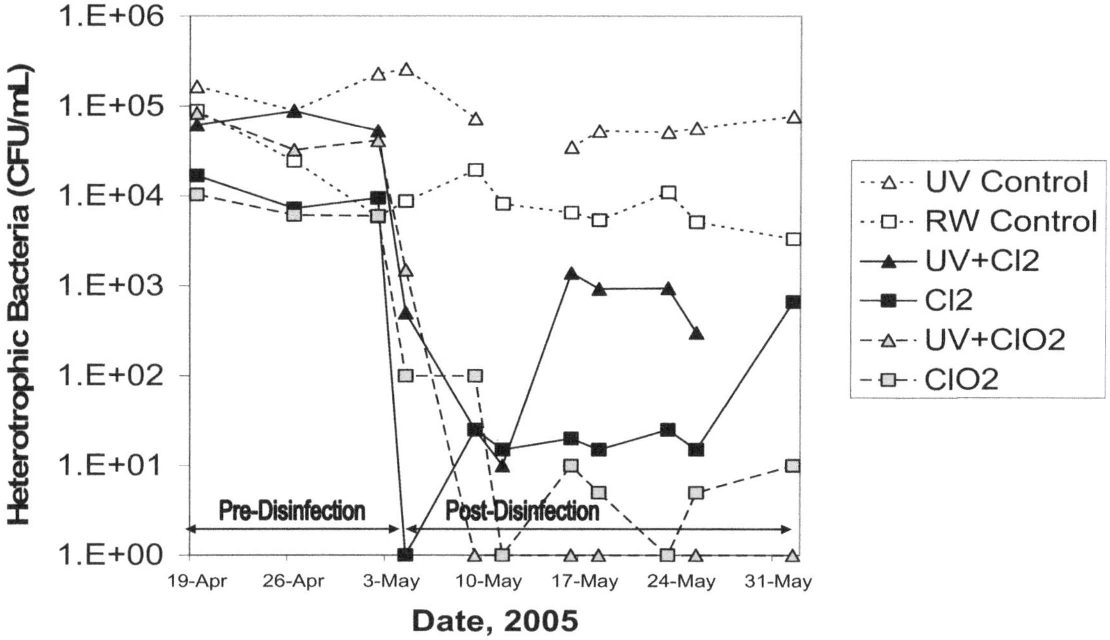

Figure 5.1 Pockwock Suspended Heterotrophic Bacteria Counts

There was not a statistically significant decrease in attached HPC bacteria for all ARs during the disinfection period (Figure 5.2). The overall average heterotrophic bacteria counts did decrease however from $6.20x10^4 \pm 4.70x10^4$ CFU/cm^2 in all ARs to $4.31x10^4 \pm 7.46x10^4$ CFU/cm^2 once chemical disinfection began. There was a greater decrease in the average HPC bacteria in ARs treated with chemical disinfection from $6.03x10^4 \pm 5.42x10^4$ CFU/cm^2 during the acclimation phase to $2.86x10^3 \pm 9.00x10^2$ CFU/cm^2 post-disinfection. The average log reduction in ARs treated with chemical disinfection was 1.14 ± 0.51, and the AR treated with UV and Cl_2 had the highest for attached HPC bacteria with a reduction of 1.71 log. Again, ARs treated with UV prior to chemical disinfection had higher log reductions of HPC bacteria. ARs that acted as controls for the experiments had no decrease in HPC bacteria in the disinfection phase for the other ARs with the UV AR having -0.28 log reduction and the RW AR with a -0.26 log reduction. There were no statistically significant differences after disinfection between biofilm or bulk log reductions of HPC bacteria.

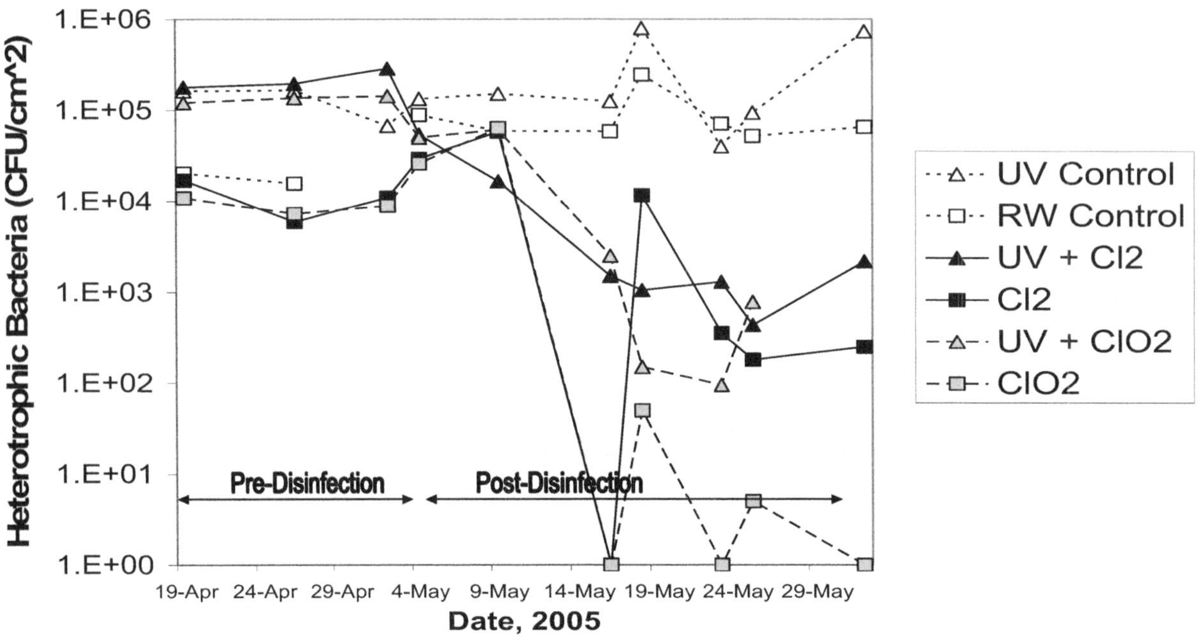

Figure 5.2 Pockwock Attached Heterotrophic Bacteria Counts

Comparison of Disinfectants

The most effective disinfection option for suspended bacteria was UV and ClO_2 (Figure 5.3). There was a statistically significant difference for bulk bacteria counts between the AR treated with ClO_2 only and AR treated with Cl_2 only, with chlorine dioxide having a higher log removal. There was also a statistically significant difference between the Cl_2 and ClO_2 ARs when pre-treated with UV, and the UV + ClO_2 AR had higher log removal than any other AR. When ARs were pre-treated with UV and then chemical disinfection there was no significant difference observed when comparing to the ARs with corresponding chemical disinfection only. All treated ARs showed significant differences when compared to the control ARs for suspended bacteria. In addition, there was a statistically significant difference between counts for the UV-treated AR and the RW AR, where the UV control had significantly higher average counts than the RW control that increased over the duration of the studies.

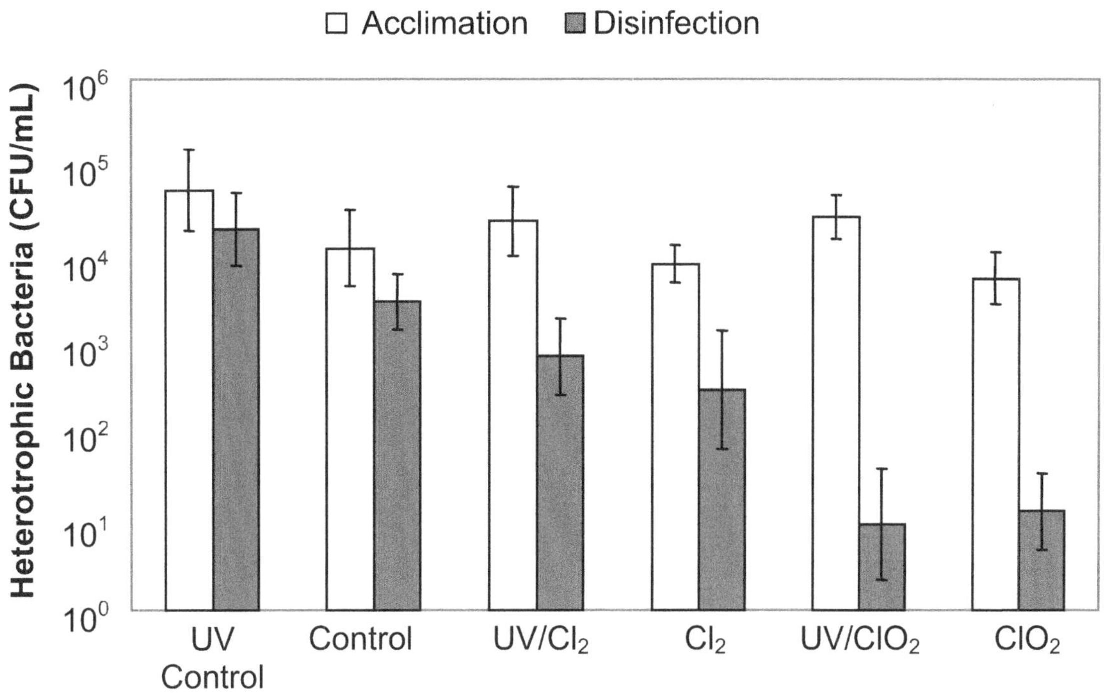

Figure 5.3 Pockwock Suspended Heterotrophic Bacteria Counts

In biofilm data the AR treated with UV and Cl_2 had the highest removal of attached bacteria with 1.71-log, however this was not statistically different than the AR treated with UV and ClO_2 with a 1.43 log removal (Figure 5.4). The AR treated with ClO_2 only had the lowest removal of attached HPC bacteria at 0.68-log, which was statistically different than the ARs treated with Cl_2 only which had an average log removal of 0.74. Each of the disinfected ARs was significantly different from the corresponding control ARs. The UV and Cl_2 combination AR was not significantly different from the Cl_2 alone AR, however the UV and ClO_2 combined AR was statistically different than the ClO_2 AR with the UV pre-treated AR having a higher log removal. In the biofilm samples, there was no significant difference between the control UV and RW ARs.

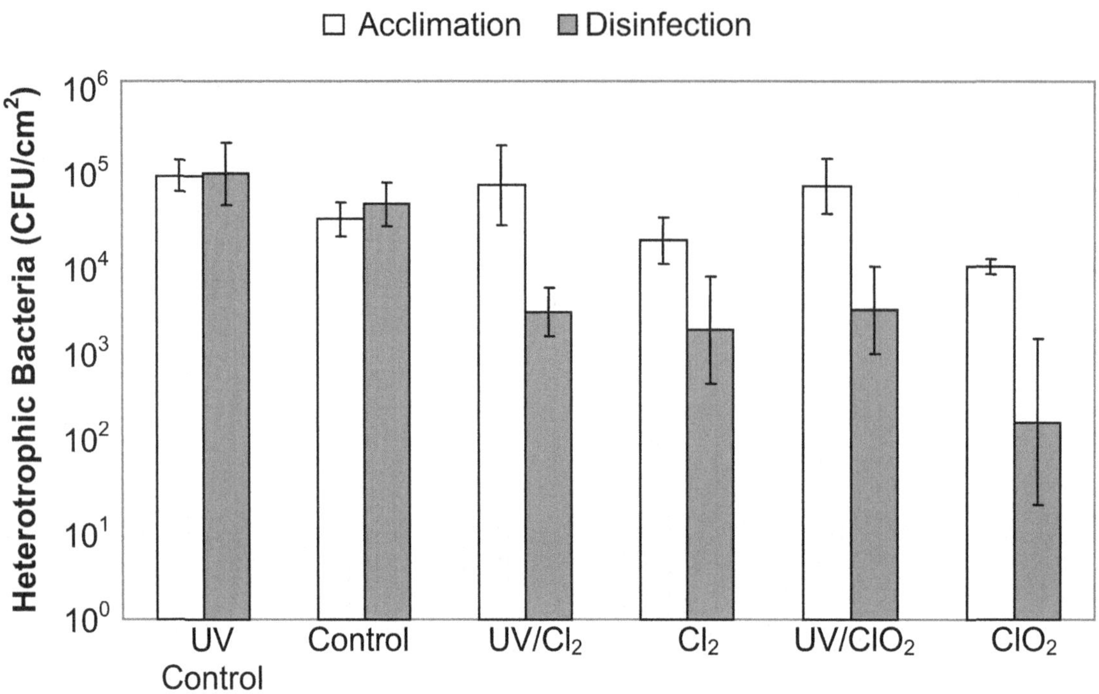

Figure 5.4 Pockwock Attached Heterotrophic Bacteria Counts

Water Quality Analysis

During the 6-week period following acclimation, ARs were dosed with chemical disinfection excluding the control ARs. The presence of UV treatment did not generally affect the required chemical dose for ARs, as has been previously demonstrated by Rand et al. (2005). The average ClO_2 dose for the AR pre-treated with UV was 1.61 ± 0.13 mg/L, which was higher than the average dose for the AR treated with ClO_2 only which was 1.52 ± 0.10 mg/L. ARs treated with Cl_2 required similar dosages to the ClO_2 ARs. The average dose for ARs treated with UV and Cl_2 was 1.65 ± 0.22 mg/L Cl_2 and 1.37 ± 0.10 mg/L for ARs treated with Cl_2 only. The resulting average chemical residual was slightly lower in ARs treated with ClO_2 than Cl_2. During the course of disinfection, the AR treated with UV and ClO_2 had an average residual of 0.17 ± 0.08 mg/L ClO_2, the ClO_2 AR averaged 0.19 ± 0.07 mg/L ClO_2, the UV and Cl_2 ARs averaged 0.21 ± 0.14 mg/L Cl_2, and finally the ARs treated with Cl_2 only had an average residual of 0.22 ± 0.06 mg/L Cl_2.

The average pH for all ARs during the acclimation phase and post-disinfection was 5.02 ± 0.20. The ARs operated at an average temperature of 21.6 ± 0.27 °C during the experiment, and the average influent temperature was slightly lower at 19.6 ± 0.37 °C.

The average turbidity for the raw water influent was 1.94 ± 1.14 NTU throughout the experiment which was higher then the average UV water influent at 1.81 ± 0.99 NTU. This was consistent with average turbidities of the control ARs with the UV control at 0.67 ± 0.41 NTU and the RW control AR at 1.10 ± 0.35 NTU. There were no significant difference between turbidity pre- and post-disinfection, and the average turbidity for all ARs was 1.00 ± 0.66 NTU during acclimation and disinfection stages.

Organic Content

Total organic carbon (TOC) in all ARs, including control ARs, was similar without significant changes over the course of the studies. The average TOC concentration in all ARs during acclimation and disinfection was 3.88 ± 0.41 mg/L. The ratio of dissolved organic carbon (DOC) to TOC was high at an average of 92.5 ± 5.7%. The average DOC concentration in all ARs during acclimation and disinfection was 3.63 ± 0.23 mg/L. UV 254 was also monitored and the average in all ARs was 0.095 ± 0.007.

Disinfection By-products

Total trihalomethanes (TTHMs) were measured following the application of disinfectants. TTHM levels were highest in ARs treated with chlorine, which is consistent with several studies that have been previously conducted (Hofmann et al., 1999; Werdehoff and Singer, 1987). The average TTHM concentration in the ARs treated with Cl_2 only was 0.0186 ± 0.0046 mg/L and 0.0254 ± 0.0085 mg/L and in the ARs treated with UV and Cl_2. The United States Environmental Protection Agency (USEPA) maximum contaminant level (MCL) for TTHMs is 0.080 mg/L. No TTHMs were observed in ARs treated with chlorine dioxide or the control ARs.

Total Coliforms and Escherichia coli

Samples were taken to analyze for total coliforms and *Escherichia coli* five times over the course of the project for raw water influent, UV-treated influent and AR effluents. No positive occurrences were observed in any sample for coliforms or *E. coli*.

SUMMARY

It was found in the Halifax field study that overall chlorine-based disinfectants in combination with UV light as a primary treatment was most effective in reducing heterotrophic bacteria when compared to UV, ClO_2, and Cl_2 treatments. Maintaining a residual of approximately 0.18 mg/L ClO_2, the UV/ClO_2 combination achieved the highest reduction for suspended (2.12 log) HPC bacteria. The UV/Cl_2 combination achieved highest reduction for attached (1.71 log) bacteria, although not significantly higher than the UV/ClO_2 combination (1.64 log). This suggests that UV in combination with either Cl_2 of ClO_2 would be a viable option for treatment of surface water to maintain biofilm control. In addition, it is possible that there were synergistic effects in enhancing the control of biofilm formation and bacterial regrowth since all ARs that were pre-treated with UV light achieved higher reductions than those that received chemical disinfection only. The benefits of sequential disinfection have been shown for other chemical basd disinfectants (e.g. Koivunen et al. 2005) and more recently Dykstra et al (2006) have demonstrated biocidal benefits from the interactions of UV and chemical disinfectants. The mechanisms are poorly understood, but suspected to be highly dependent on the disinfectant. UV treatment alone was not able to control regrowth in the AR systems and in this study actually increased counts over time, potentially due to fouling within the lamp casing.

CHAPTER 6
FIELD SCALE DISINFECTION STUDIES – GROUNDWATER

INTRODUCTION

Treatment of drinking water in warm climates can be difficult considering degradation of disinfection residual in finished water storage facilities and distribution systems that are sometimes subject to high temperatures. Chloramines are often considered as an alternative to chlorine since they have been shown to persist longer in a distribution system (Norman et al., 1980). Martel et al. (2002) found that chloramines improved water quality in storage facilities in a warm climate where chlorination resulted in residual loss and increased bacteria counts. However, when considering a switch to chloramines, utilities must consider other possible issues that may arise in warm climates, primarily the potential for nitrification. It has been reported in literature that conditions that favour nitrification include warm climates or high temperatures in summer months (McGuire et al., 2006; Pintar and Slawson, 2003). Pryor et al. (2004) found that a switchover to monochloramine at a utility in Florida resulted in lower Legionella but increased total coliforms and HPC bacteria and also caused nitrification. Nitrification leads to decreased residual and increased bacteria levels in distribution systems. In addition, nitrate and nitrite concentrations increase during nitrification and these DBPs may lead to serious illnesses such as blue-baby syndrome and cancer.

There have been few studies focused on disinfection of drinking water in distribution systems conducted in warm climates. Most annular reactor experiments have been carried out in cooler climates (Gagnon et al., 2005; Batte et al., 2003; Sharp et al., 2001; Camper, 1996). With changing standards that are becoming stricter on DBP formation, it is estimated that several more utilities will consider changing to chloramines from chlorine to lower THM and HAA formation. It is important for these utilities to consider benefits and implications for the distribution system when switching disinfectants.

There is also the possibility for enhanced disinfection through combination strategies, since previous studies have shown synergistic benefits with sequential disinfection (Dykstra et al., 2006; Koivunen et al., 2005; etc.). Investigation of implementing UV light in combination with chlorine-based disinfection is needed to determine the possibility of enhanced removal of bacteria in warm climate distribution systems. A potential concern related to the implementation of UV treatment is the interaction between UV light and chemical disinfection. Some studies have shown that free chlorine or monochloramine in water exerts demand on the UV light through absorbance of UV irradiation, and conversely that the UV light destroys the chemical disinfectants reducing the available residual for inactivation of microbial cells. Ormeci et al. (2005) found that although UV absorbance of free chlorine and monochloramine was relatively small, they may affect the effectiveness of UV light toward targeted microorganisms. This study also found the chlorine and monochloramine in potable water decay steadily in presence of UV light, especially chlorine in poorer water quality.

The focus of this chapter was to determine the effectiveness of Cl_2, NH_2Cl, UV/Cl_2, and UV/NH_2Cl in suppressing biofilm growth and controlling microbial pathogens in drinking water distribution systems through a field-scale study, and to compare treatment with UV to no UV treatment. Heterotrophic plate counts and coliforms were used to compare disinfection treatments for bacteria within the systems. In addition, water quality parameters were monitored throughout the experiments. A second objective was to determine the extent of free chlorine and monochloramine decay in the presence of UV light, and the potential for these chemicals to absorb UV light. Observations from the field study were analyzed to meet this objective in addition to a separate bench-scale study.

RESULTS AND DISCUSSION

Unlike the other field experiments presented, the annular reactors for this study went through no acclimation period and therefore average counts include every sample throughout the 30-week study. Samples for HPC bacteria included the influent to each AR, the effluent of each AR as well as biofilm samples from the coupons of each AR. The overall mean number of suspended heterotrophic bacteria in the influents was $6.10\times10^5 \pm 4.81\times10^5$ CFU/mL and $4.04\times10^5 \pm 5.11\times10^5$ CFU/mL in the effluent of all ARs for the duration of the project. The mean attached (biofilm) heterotrophic bacteria for all ARs was higher at $1.79\times10^6 \pm 8.01\times10^5$ CFU/cm^2. The average counts for both bulk samples were statistically different than biofilm average counts but were not significantly different from each other.

Comparison of Disinfectants

The most effective disinfection option for influent suspended bacteria (Figure 6.1) was Cl_2 alone with an average count over the course of the experiment of $4.20\times10^2 \pm 1.37\times10^3$ CFU/mL which was a statistically significant difference compared to all other ARs. The highest average count was observed in the AR treated with NH_2Cl alone at $1.09\times10^6 \pm 1.17\times10^6$ CFU/mL but this did not differ significantly from the UV + NH_2Cl AR with an average count of $8.33\times10^5 \pm 7.17\times10^5$ CFU/mL.

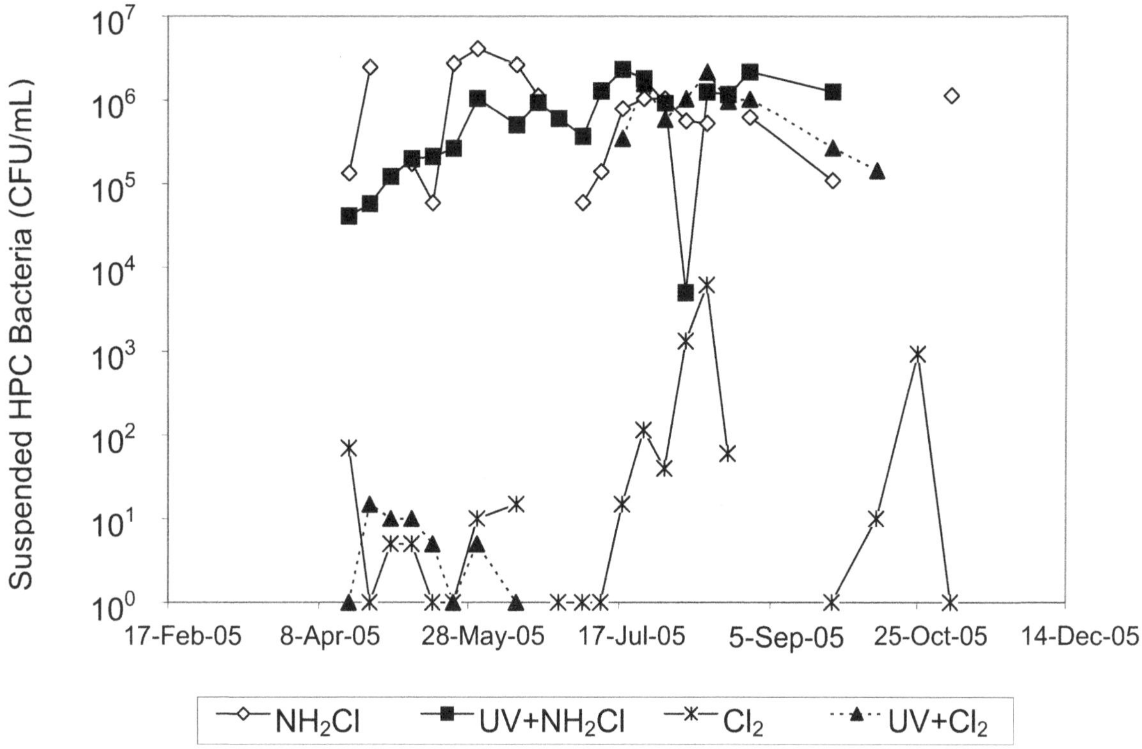

Figure 6.1 Average Suspended Heterotrophic Bacteria in the Influent Streams

The suspended heterotrophic bacteria counts increased over the course of the experiment in the AR treated with UV + Cl_2 jumping from an average count of $6.00 \times 10^0 \pm 5.21 \times 10^0$ CFU/mL in the first eight weeks of the project to $8.90 \times 10^5 \pm 6.59 \times 10^5$ CFU/mL for the final 20 weeks resulting in an overall average of $5.15 \times 10^5 \pm 6.67 \times 10^5$ CFU/mL. The increase in heterotrophic bacteria for the UV and Cl_2 AR corresponded to decrease in free chlorine in the influent to below detection limits (Figure 6.2). A potential explanation would be that demand in the UV influent line was exceeded due to increased biofilm formation and degradation of chlorine in the UV lamp. This decrease in residual did not occur in the AR receiving water treated with Cl_2 only, where although data is scattered as is common in field studies, overall residual and heterotrophic bacteria concentrations remain fairly constant overall (Figure 6.3). Nevertheless this data set indicated that maintaining chlorine residual was essential for the control of heterotrophic bacteria and is consistent with other reports (LeChevallier et al., 1996; Huck and Gagnon, 2004; Baribeau et al., 2005).

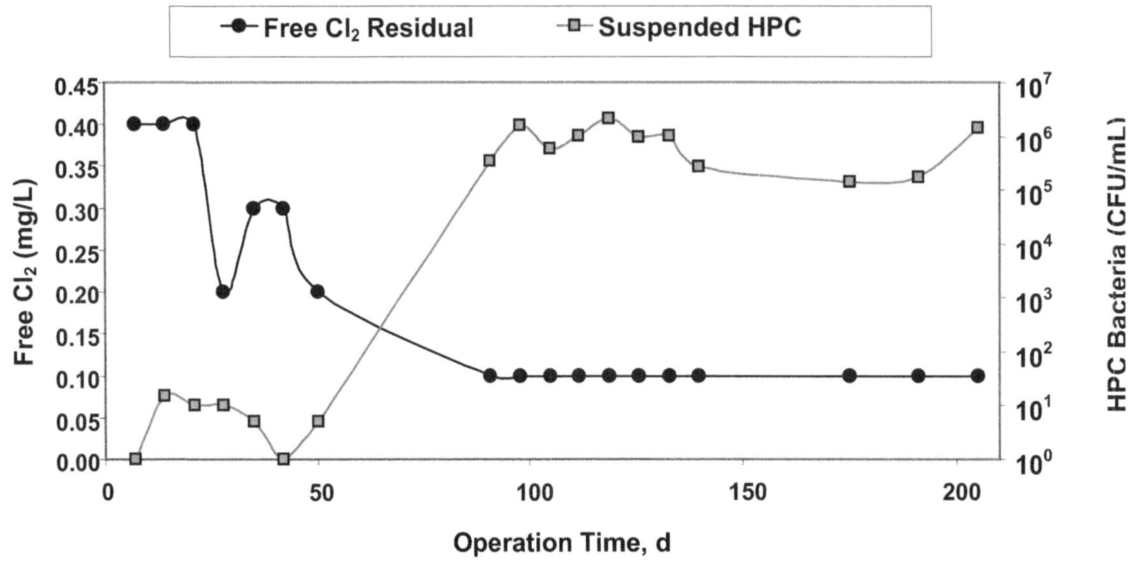

Figure 6.2 Free Chlorine Residual with Corresponding HPC Bacteria in the Influent of the AR Receiving Water Treated with UV and Cl_2

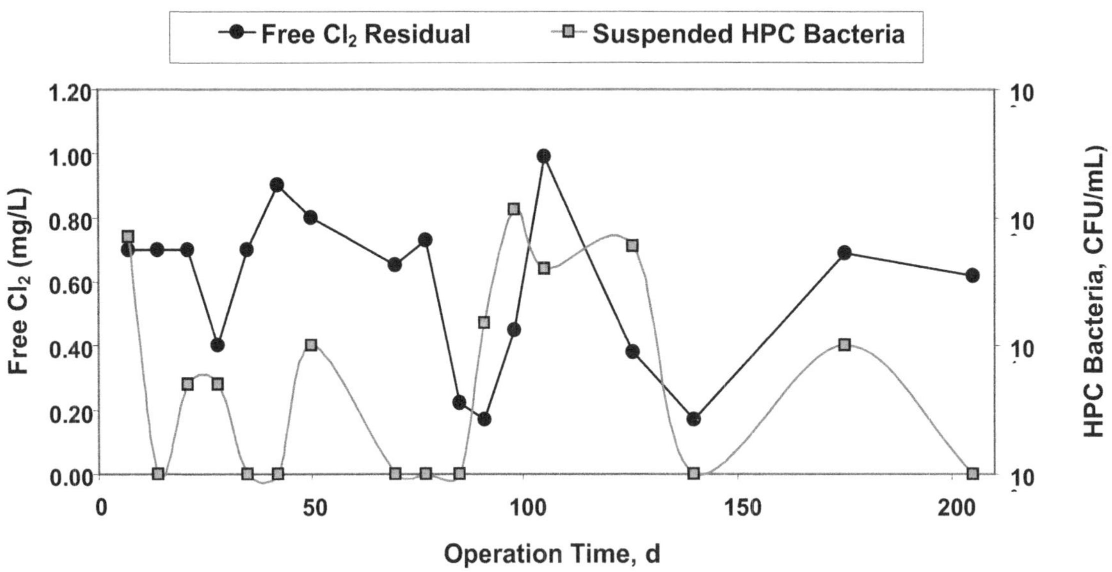

Figure 6.3 Free Chlorine Residual with Corresponding HPC Bacteria in the Influent of the AR Receiving Water Treated with Cl_2 Only

Similarly to influent data, the AR treated with Cl_2 alone had the lowest average count for effluent samples at $8.19x10^2 \pm 9.38x10^2$ CFU/mL and differed significantly from all other ARs (Figure 6.4). The UV + Cl_2 AR had increasing counts in effluent samples as was observed in its influent however not as significant going from $3.60x10^4 \pm 4.83x10^4$ CFU/mL to $6.31x10^5 \pm 4.76x10^5$ CFU/mL. This resulted in an overall average of $3.53x10^5 \pm 4.40x10^5$ CFU/mL which did not differ significantly from the AR treated with UV + NH_2Cl or NH_2Cl alone. The highest average count was observed in the UV + NH_2Cl AR at $1.00x10^6 \pm 1.19x10^6$ CFU/mL which was significantly different than the NH_2Cl AR which had an average count of $2.61x10^5 \pm 2.51x10^5$ CFU/mL. Overall for bulk samples of the effluent, ARs treated with UV in addition to the chemical disinfectants had higher average counts compared to ARs with no UV treatment.

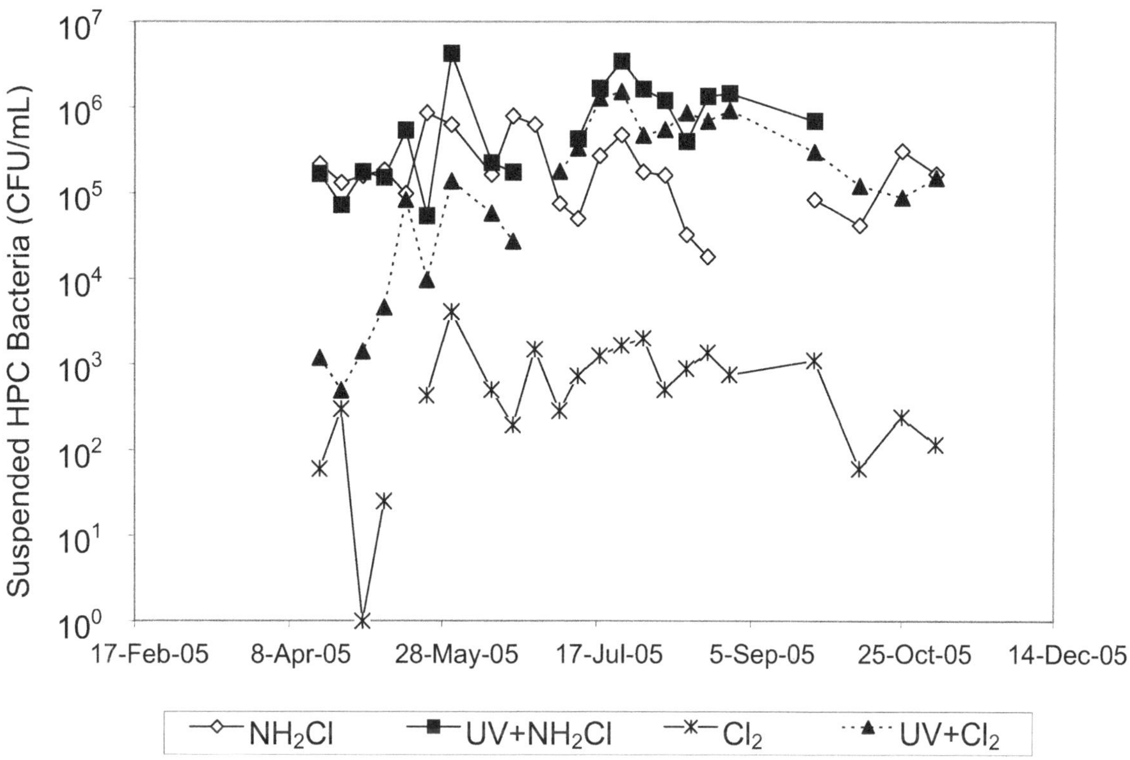

Figure 6.4 Average Suspended Heterotrophic Bacteria in the Effluent of Each AR

In biofilm data the AR treated with Cl_2 had the lowest level of attached bacteria with an average count of $4.11x10^5 \pm 7.38x10^5$ CFU/cm^2 (Figure 6.5). This was a significantly lower count compared to all other ARs, but there was no statistically significant difference between average counts for the remaining three ARs. The AR treated with UV and Cl_2 had the highest average count overall at $2.49x10^6 \pm 2.43x10^6$ CFU/cm^2. The AR that received NH_2Cl treated water had an average count of $1.86x10^6 \pm 2.40x10^6$ CFU/cm^2 and the UV + NH_2Cl AR averaged $2.41x10^6 \pm 2.11x10^6$ CFU/cm^2 for attached heterotrophic bacteria. Similarly to effluent samples, ARs with the additional treatment of UV had higher overall average counts than the ARs that had no UV treatment.

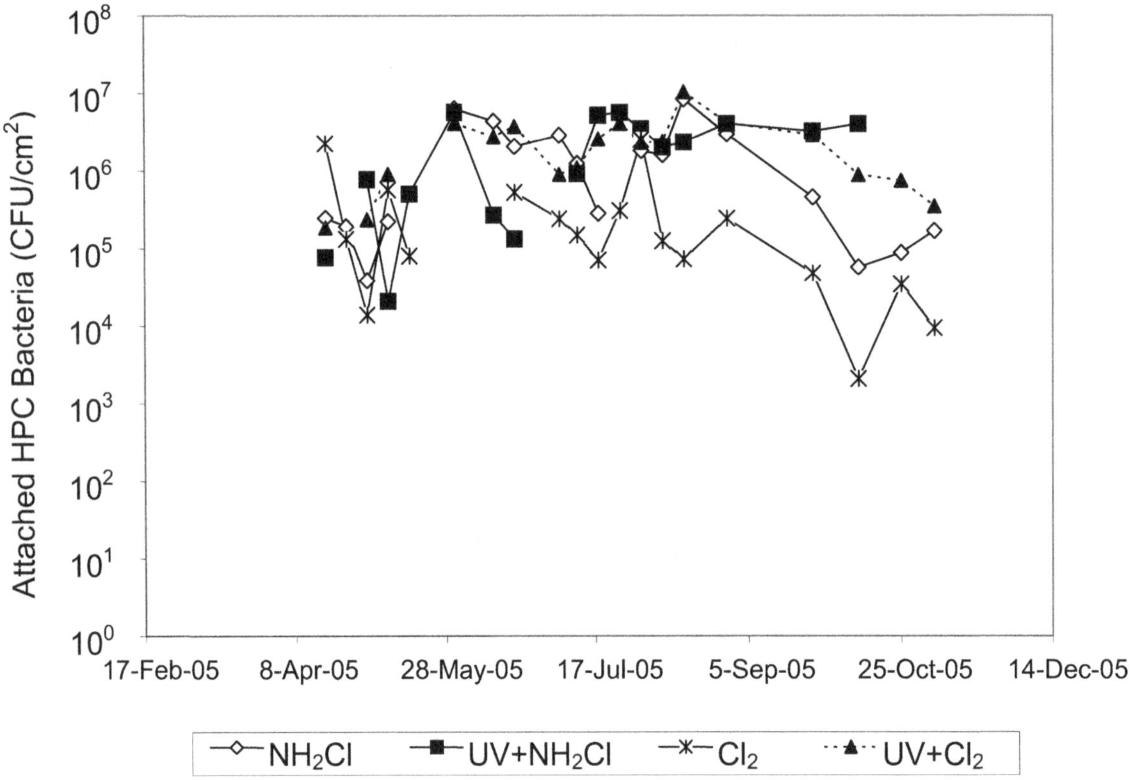

Figure 6.5 Average Attached Heterotrophic Bacteria from PVC Coupons

Water Quality Analysis

Over the 30-week study period water samples were taken of influent and effluent water streams to analyze for several parameters including pH, temperature, turbidity, TOC, disinfectant residual, total coliforms, and *Escherichia coli*. In addition, samples were monitored both onsite and in a laboratory for ammonia and nitrates. Finally samples were taken to analyze from the disinfection by-products of TTHMs three times over the course of the experiment.

Disinfectant Residual

For both chlorine and monochloramine, the presence of UV treatment appeared to decrease disinfectant residuals in both influent and effluent water streams. The average residual NH_2Cl in the influent of that AR was 1.67 ± 0.74 mg/L compared to 0.90 ± 0.66 mg/L for the UV + NH_2Cl AR influent. Similarly, the NH_2Cl residual was lower in the UV + NH_2Cl effluent with an average of 0.62 ± 0.65 mg/L compared to 0.87 ± 0.60 mg/L in the effluent of the AR receiving water treated only with NH_2Cl.

This trend held true free and total chlorine residuals in influent and effluent samples. The average influent free Cl_2 residual for the AR also treated with UV was 0.17 ± 0.11 mg/L compared to 0.53 ± 0.26 mg/L in the Cl_2 alone AR. Also, the average effluent residual concentration was 0.11 ± 0.03 mg/L in the UV + Cl_2 AR and 0.32 ± 0.17 mg/L in the Cl_2 AR.

As previously mentioned, free chlorine residual concentration in the UV + Cl_2 AR influent and effluent streams was below the detection limit of 0.10 mg/L for the majority of the experiment. Total chlorine levels in the influent streams were also lover in ARs treated with UV (UV + Cl_2 = 0.27 ± 0.10 mg/L, Cl_2 = 0.68 ± 0.32 mg/L, UV + NH_2Cl = 1.30 ± 1.16 mg/L, NH_2Cl = 2.13 ± 0.80 mg/L), which was also observed in effluent streams (UV + Cl_2 = 0.20 ± 0.02 mg/L, Cl_2 = 0.43 ± 0.18 mg/L, UV + NH_2Cl = 0.86 ± 0.86 mg/L, NH_2Cl = 1.33 ± 0.83 mg/L).

pH, Temperature and Turbidity

The average pH for all ARs in the influent water streams was 7.76 ± 0.04 and 7.84 ± 0.02 in effluent streams. The ARs operated at an average temperature of 25.2 ± 0.16 °C during the experiment, and the average influent temperature was slightly higher at 26.3 ± 0.23 °C. In comparison, during the field tests in Halifax, PW and EBMUD the mean influent temp was 19.4 ± 2.79 °C, 12.7 ± 2.77 °C and 24.0 ± 2.75 °C respectively.

The average turbidity for the NH_2Cl influent water streams was higher than the Cl_2 influent water streams. In addition, water treated with UV had lower turbidity on average than water that was not treated with UV. The average turbidity for the UV + NH_2Cl influent was 1.58 ± 1.75 NTU and 2.12 ± 3.45 NTU for the NH_2Cl influent. For effluent samples, the average turbidity in the UV + NH_2Cl AR was 0.73 ± 0.35 NTU and 0.89 ± 0.81 NTU in the NH_2Cl AR. Average influent turbidities for the UV + Cl_2 and Cl_2 ARs were 0.45 ± 0.29 NTU and 0.50 ± 0.26 NTU respectively, and for effluent the average turbidity for the UV + Cl_2 and was 0.31 ± 0.12 NTU and 0.38 ± 0.21 NTU for the Cl_2 AR.

Organic Content

Total organic carbon (TOC) was similar without significant changes over the course of the study and the presence of UV light did not appear to affect TOC levels. The average influent TOC concentration in all ARs was 3.81 ± 0.08 mg/L and 3.92 ± 0.20 mg/L in the effluent of all ARs.

Disinfection By-products

Total trihalomethanes (TTHMs) were measured three times near the end of the experiment in September and October 2005. The AR receiving NH_2Cl and UV treated water was shut down in October due to operational problems therefore only one TTHM measurement was obtained for that AR. On average, ARs treated with UV had lower TTHM levels and chlorinated samples had higher levels. The average TTHM concentration in the AR treated with Cl_2 alone was 53.7 ± 5.3 μg/L and 43.3 ± 3.2 μg/L in the AR treated with UV and Cl_2. The measurement obtained for the UV + NH_2Cl AR showed a TTHM concentration of 41.6 μg/L and the AR treated with NH_2Cl alone had an average TTHM concentration of 48.5 ± 2.0 μg/L. The United States Environmental Protection Agency (USEPA) maximum contaminant level (MCL) for TTHMs is 0.080 mg/L.

Total Coliforms and Escherichia coli

Data was obtained for total coliforms and *E. coli* six times over the course of the project. For influent samples, the water treated with UV and NH_2Cl was positive for total coliforms once resulting in an average concentration of 1.03 MPN/100mL, and the NH_2Cl treated water was positive twice resulting in an average concentration of 0.87 MPN/100mL. These occurrences were negative for *E. coli*. The chlorinated influent water streams had no positive occurrences for total coliforms. There were positive occurrences in all AR effluents. The effluent for the NH_2Cl and UV + NH_2Cl ARs each tested positive five out of six times resulting in average concentrations of 11.8 MPN/100mL and 5.98 MPN/100mL respectively. The Cl_2 AR tested positive three out of six times resulting in an average concentration for total coliforms of 6.77 MPN/100mL. The UV + Cl_2 had positive occurrences 4 times which resulted in an average concentration of 6.75 MPN/100mL. No occurrences in any samples tested positive for *E. coli* throughout the experiment.

Nitrate and Nitrite

Nitrate and nitrite were measured in the Pinellas County laboratory for the entire study period. Nitrification was suspected to occur in water streams treated with NH_2Cl only. Samples collected from the Cl_2 streams and corresponding chlorinated ARs had nitrite and nitrate concentrations that were below the detection limit for all sample dates. The average nitrate concentration for the NH_2Cl influent stream measured in the lab was 0.07 ± 0.05 mg/L as N and 0.09 ± 0.05 mg/L as N for the UV + NH_2Cl influent. The effluent of the NH_2Cl AR had an average nitrate concentration of 0.12 ± 0.07 mg/L as N and the average concentration was 0.30 ± 0.34 mg/L as N for the effluent of the UV + NH_2Cl AR in lab analysis.

Nitrite concentrations were also measured for AR influent and effluent streams. The influent of the NH_2Cl AR had an average nitrite concentration of 0.12 ± 0.10 mg/L as N in lab analysis and the UV + NH_2Cl stream averaged 0.19 ± 0.15 mg/L as N. The effluent of the NH_2Cl AR averaged 0.18 ± 0.15 mg/L as N and the UV + NH_2Cl AR had an average nitrite concentration of 0.13 ± 0.19 mg/L as N in lab analysis of the effluent stream. Observing the data, the AR receiving UV-treated water had higher concentration levels of nitrate and nitrite compared to the AR with NH_2Cl disinfection only, indicating photolysis of NH_2Cl may have occurred.

Free and Total Ammonia

Free ammonia for water streams treated with NH_2Cl was measured at the Pinellas County laboratory and in addition in the field from June 2005 to the end of the experiment as a check for lab analysis. Field and lab results were fairly consistent therefore only lab analysis results are reported. The average free ammonia concentration in the influent treated with NH_2Cl alone was 0.44 ± 0.14 mg/L as N, and for the UV + NH_2Cl influent the average was 0.46 ± 0.15 mg/L as N in the lab analysis.

Total ammonia was measured in the lab alone for all water streams including the chlorinated ARs. Average influent concentrations were slightly higher than effluent concentrations and ARs that were not treated with UV had slightly lower total ammonia levels. The average influent concentrations for the NH_2Cl and the UV + NH_2Cl were 0.89 ± 0.26 mg/L as N and 0.69 ± 0.30 mg/L as N respectively. The effluent of the NH_2Cl AR had an average concentration of 0.73 ± 0.32 mg/L as N and the average was 0.50 ± 0.41 mg/L as N for the UV + NH_2Cl AR. The Cl_2 ARs all measured close to the detection limit for free and total ammonia.

Free ammonia, nitrite and nitrate concentrations were all measured as nitrogen for analysis, and in order to determine if nitrification occurred in the ARs treated with monochloramine, the variation in concentration of each in the effluent over the course of the experiment was analyzed. Figure 6.6 shows that free ammonia, nitrate and nitrite concentration in the NH_2Cl AR varied but did not change overall during the study, indicating nitrification didn't occur in this AR.

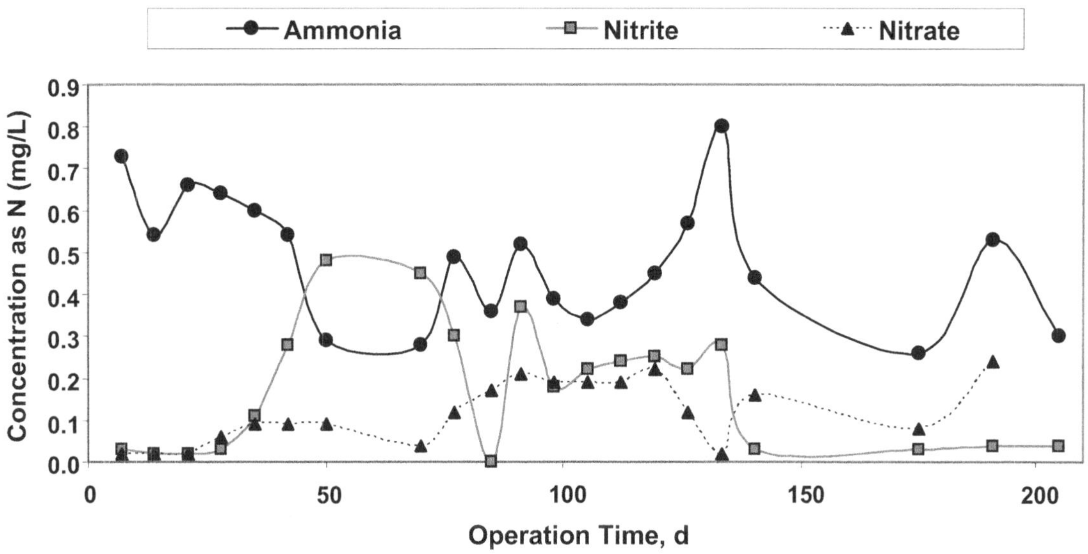

Figure 6.6 Ammonia, Nitrite and Nitrate Concentrations in the Effluent of the AR Treated with NH_2Cl Only

Nitrification did appear to occur in the AR receiving NH_2Cl and UV-treated water (Figure 6.7). Nitrification is a process that can take up to several weeks in a distribution system and occurs when free ammonia is oxidized into nitrite which is additionally oxidized to form nitrate (McGuire et al., 2006). From Figure 6.7 it can be observed that a reduction in free ammonia concentration occurred in the 12th week of the study, which corresponded to an increase in nitrite concentration. This was followed by a decrease in nitrite levels and a corresponding increase of nitrate concentration in the 16th week. Accelerated nitrification in systems with UV light disinfection following monochloramine has not yet been reported in literature. Ammonia (NH_3) concentrations in the first 11 weeks of the experiment were similar in both ARs (NH_2Cl = 0.50 ± 0.15 mg/L as N and UV+NH_2Cl = 0.55 ± 0.16 mg/L as N), therefore nitrification didn't occur in the UV + NH_2Cl AR due to a higher level of excess NH_3 compared to the NH_2Cl AR. A possible explanation could be similar to that for increased HPC counts in Cl_2 and NH_2Cl ARs with UV treatment, where UV light increases bacterial regrowth potential of NOM (Parkinson et al., 2003), thereby causing elevated levels of ammonia oxidizing bacteria (AOB) which leads to oxidation of ammonia into nitrite. In addition, Mack and Bolton (1999) established –OH radicals are the result of photolysis of nitrate and nitrite, which could be achieved with the UV lamp. Gagnon et al. (2004) suggest -OH radicals react with humic material in water to form biodegradable organic matter, which act as a nutrient source for bacteria in distribution systems. It is possible that AOB used the BOM as a source and were able to thrive and reproduce as part of the biofilm community in the UV-treated AR, which eventually led to occurrence of nitrification. Also, photolysis caused by UV light would have increased degradation of NH_2Cl and released more ammonia, which would also allow AOB to proliferate (McGuire et al., 2006). Without the elevated levels of AOB in the AR receiving water treated with NH_2Cl nitrification was not accomplished.

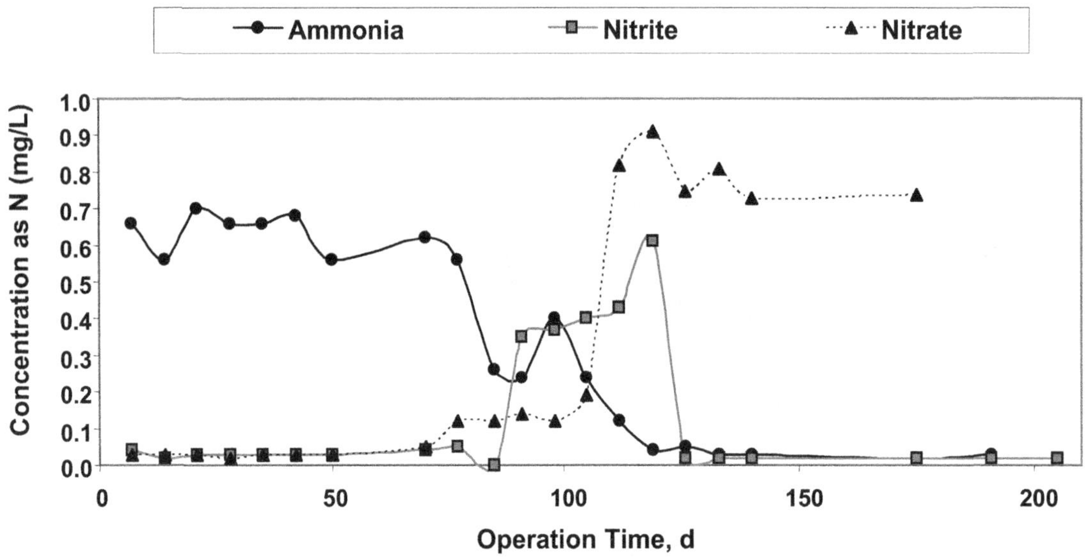

Figure 6.7 Ammonia, Nitrite and Nitrate Concentrations in the Effluent of the AR Treated with UV and NH_2Cl

SUMMARY

Treatment with chlorine alone was the most effective treatment in the Florida study with lower average counts for influent, effluent and biofilm samples. Monochloramine was not effective in controlling HPC bacteria growth and there were several occurrences where bacteria within the tubing system hindered flow rates. It was found that the addition of UV treatment hindered control of HPC bacteria. A separate study confirmed concerns that UV light was degrading chemical residual therefore there was less free chlorine and monochloramine within the systems to reduce bacteria.

CHAPTER 7
PROJECT SUMMARY AND RECOMMENDATIONS

PROJECT SUMMARY

The purpose of this chapter is to summarize the findings presented in this reports in the previous chapters. Overall, this report has examined the occurrence and mitigation of *E. coli* present in distribution systems.

UV treatment was extremely effective at eradicating *E. coli* prior to entering the model distribution system. In fact, synergistic effects for mitigating *E. coli* were observed when *E. coli* was spiked into the UV-reactor and then was treated with either ClO_2 or Cl_2. the benefits of combined disinfectants were only observed on *E. coli* K12, as the mut mutant was completely inactivated by UV-treatment. In fact, the mut mutant was not recovered from any AR following an initial UV treatment. It was also found that *E. coli K12* was much more resistant to UV disinfection that the mutS mutant strain. The mutant strain was never cultured from the biofilm or effluent of any of the Post-UV ARs.

Results from the lab study and field studies were also consistent with findings report by a previous AwwaRF study (Gagnon et al. 2004). In particular, a synergy between UV and chlorine as well as chlorine dioxide in reducing heterotrophic bacteria using ARs was observed in both projects. Interestingly, UV carried no added benefit for inactivating *E. coli* in the AR, for ARs that were spiked following UV treatment. This finding confirms that UV treatment has a positive impact on pathogen control at the plant that can be extened into the distribution system, but does not modify the water matrix or mircoflora enough to prevent subsequent regrowth from pathogen intrusion.

Overall, phenotypical responses of *E. coli* isolated from the AR biofilms varied based on pipe material, UV disinfection and chemical disinfectant. It was concluded that there are several explanations to why *E. coli* isolated from the annular reactor biofilms differed in phenotypical profile (API 20E test) from the original stock culture used to simulate a breech in distribution system integrity. Some factors that may have influenced these phenotypical changes include: diversification of organisms within the biofilm communities and sublethal injury as a result of chemical disinfection. In addition, UV treatment and pipe material may have influenced them to change phenotypically in order to survive and cope in a different environment. As a result, they may have developed a "biofilm phenotype" which varied from that of the stock and planktonic culture (Parsek and Fuqua 2004).

Additional molecular and antimicrobial resistence analysis was conducted on identifying bacterial species in the ARs, as well as determined antimicrobial resistance. It was shown that different species were present in each AR when exposed to monochloramine, free chlorine, and chlorine dioxide. In particular, DGGE only confirmed the presence of *E. coli* in the cast iron chlorine dioxide trials.

An additional objective of the project was to test the theory of synergy outside of a controlled setting over a longer term. Two studies employed ARs in the field to compare disinfection strategies for surface water, and surface and groundwater blended water. It was found in the surface water studies that combination disinfection strategies achieved the highest log removals of attached and suspended heterotrophic bacteria. When calculations for synergy were applied they were positive for attached and suspended bacteria in one study but only attached for a second surface water study. It was found in this study that raw water quality improved over the course of the project, allowing for high log reduction in the UV control AR, which led to a negative calculated synergy result. However, the combination strategies achieved higher log reduction for the suspended bacteria, indicating UV light damages cells to make them more susceptible to secondary disinfection. When comparing data from each field study to the laboratory study, it was observed that results were typically more scattered in the field. The changing water quality in field studies made simple synergy analysis less applicable when compared to a controlled laboratory setting. However, it was also observed that although water matrix caused variation in the effectiveness of disinfectants, combination strategies consistently achieved higher inactivation of HPC bacteria in bulk and biofilm samples.

Considering data from surface water field study, it can be concluded that the implementation of UV light as a primary disinfectant enhanced removal of attached and suspended heterotrophic bacteria and results indicated that synergy exists between UV light and chlorine and chlorine dioxide. UV was not effective when used as the only disinfectant, and increased counts were observed in the UV ARs of the Halifax study. This occurrence could be accredited to several possibilities, including higher re-growth potential of NOM as was observed by Parkinson et al., (2003) or fouling in the UV experimental system. These issues would result in higher counts when no residual protection is supplied, which would lead to increased biofilm formation, as was seen in most UV ARs in the studies. However, when chemical disinfection was applied following the acclimation period, the synergistic benefits between UV and the secondary disinfectants decreased viable cells entering the systems and also applied residual which overall was able to reduce suspended and biofilm bacteria.

In the Florida study, when UV light followed disinfection with chlorine or monochloramine, suspended and attached HPC bacteria generally increased compared to treatment with chemical disinfection alone. This contradicted findings from the laboratory that concluded pre-chlorination followed by UV disinfection improved microbial control and inhibited organic material deposits on the quartz tube of the lamp itself. However, this study also recommended the use of secondary disinfection with Cl_2 and did not consider by-product formation.

It was observed in the Florida field study that residual concentrations were reduced in ARs treated with UV light post-chemical disinfection and it was hypothesized that UV light degraded chlorine and monochloramine. A sudden increase in suspended bacteria in the influent of the UV and Cl_2 AR directly corresponded to a drop in Cl_2 residual below detection limits, most likely due to increased demand and degradation of Cl_2 by UV light. Ormeci et al. (2005) found chlorine and monochloramine decayed at several different dosages of UV irradiation. It is believed that in the Florida study UV irradiation caused the photolysis of chlorine and monochloramine. Nowell and Hoigne (1992) found that the photolysis of Cl_2 in sunlight occurred with a half life of 12 minutes and suggested UV photo-oxidizes chlorine into the chloride ion. Degradation of chemical concentration was confirmed in a controlled laboratory where it was found that chlorine, monochloramine and chlorine dioxide decreased up to 66% with exposure to UV light compared to controls, especially in poorer quality water. Furthermore, there was a loss of Cl_2 residual during the project following UV treatment. This loss of residual substanticially increased growth in the ARs, which confirms that the point of application is critical for UV treatment. It was also observed that UV treatment as a secondary disinfectant promoted microbial regrowth and nitrification in chloraminated systems most likely because it degraded residual concentration and indirectly provided increased nutrient sources for heterotrophic and ammonia-oxidizing bacteria.

Considering all observations from the two pilot studies and the laboratory study, it is concluded that the sequential inactivation of heterotrophic bacteria with UV light and secondary chlorine-based disinfectants shows synergistic benefits. In the controlled laboratory setting, synergistic effects were clearly observed for the inactivation of *E. coli*. In comparing disinfection strategies in field studies, combination treatments with UV as a primary disinfectant generally achieved highest log removals, and chlorine dioxide was the most effective disinfectant followed by chlorine and finally monochloramine. UV alone was not effective against heterotrophic bacteria and possibly increased bacterial regrowth potential in the ARs which let to increased HPC bacteria without secondary residual disinfection. Enhanced reduction with UV light in combination with secondary chlorine-based disinfection was observed in all water types regardless of quality or climate.

RECOMMENDATION FOR THE WATER INDUSTRY

It is recommended that implementing UV light as a primary disinfectant with chlorine-based secondary disinfection would enhance removal of HPC bacteria due to synergistic effects and the previously established added benefit of a wider range of bacteria and protozoa inactivated (i.e. chlorine-resistant pathogens such as *Cryptosporidium parvum*). Utilities would be able to benefit from high reductions in HPC bacteria and lower formation of biofilm in the distribution systems. There would also be potential for lower CTs required for chlorine-based disinfectants with UV light pre-treatment to achieve higher reductions than chemical disinfectant alone at high Ct. However, utilities would not be able to depend on UV pre-treatment to lower required dosages of chlorine-based disinfectant to maintain minimum residual concentrations. It is also recommended that UV light not be used as the primary and only disinfection because bacteria counts actually increased with UV-treated water when no residual protection was supplied.

To achieve downstream benefits of UV treatment, it is recommended that UV treatment precedes as a secondary disinfectant. If UV follows chemical disinfection, the degradation of chemical residual in the presence of UV light produces negative results. Although studies have shown that pre-chlorination aids in controlling biological growth on the lamp itself, it is believed that this benefit is overshadowed by the negative impacts of pre-chemical disinfection (i.e. loss of Cl_2 residual). The reduction of residual due to UV treatment leads to increased bacteria counts because there is less available to control microbial regrowth in the distribution system. To combat this problem, utilities would need to additionally treat with chemical disinfection following UV irradiation in order to increase residual protection, which would potentially lead to higher disinfection by-product formation and increase chemical costs for disinfection. It is especially not recommended to implement secondary UV disinfection with monochloramine as a primary disinfectant, since it has been observed that nitrification occurred with this disinfection strategy leading to higher nitrate and nitrite concentrations.

The laboratory study presented in this project showed synergistic effects in reducing *Escherichia coli*, which is a pathogen easily reduced by UV light and chemical disinfection. It would be recommended to investigate if synergistic benefits still occur with a pathogen that would not be easily affected by UV light, and to determine if multiple damage mechanisms still enhance removal.

Furthermore, this project and other studies have shown that disinfection synergy occurs with sequential disinfection strategies. However, the mechanisms involved have not been identified. Molecular work in the present study indicates potential pathways for exploring the benefits of sequential treatment. However, further research is needed to more fully characterize possible mechanisms of synergy.

While neither knowledge of the biofilm composition nor the presence of antibiotic resistant mutants are likely to have immediate impacts on water treatment, It should be recognized that such information allows a better understanding of what is occurring downstream of the treatment plant.Biofilm composition is likely to be somewhat site specific and source water dependent as well as being affected by pipe materials and upstream treatments. It may also be affected at certain points in the distribution system by breakages, pressure transients, etc. As knowledge in this area accumulates it is probable that the presence of certain species may be able to predictive of events such as nitrification, leakage etc.

Antibiotic resistance is a widespread phenomenon and may not have more significance for drinking water than for other areas where selective pressures are applied. Nevertheless, the longer that pathogens are present in a distribution system the greater is the opportunity for potential resistance to occur. If there is a risk it would therefore likely be greatest at the ends of the distribution system or in low flow or dead end areas. Those at most risk are individuals immunocompromized through age, disease or medication, but they are also at greatest risk from any pathogens

REFERENCES

American Public Health Association (APHA). 1995. Standard Methods for the Examination of Water and Wastewater, 19th ed. Washington, DC

Armstrong, JL, Calomiris JJ,and Seidler RJ 1982. Selection of antibiotic-resistant standard plate count bacteria during water treatment. *Appl. Environ. Microbiol.* 44: 308-316.

Armstrong, JL, Shigeno, DS, Calomiris JJ,and Seidler RJ 1981. Antibiotic resistant bacteria in drinking water *Appl. Environ. Microbiol*. 42: 277-283.

AWWA Water Quality Division Disinfection Systems Committee 2000. Committee Report: Disinfection at Large and Medium-Size Systems. *Journal AWWA*, 92:5:32-43.

Batte, M., Koudjonou, B., Laurent, P., Mathieu, L., Coallier, J. and Prevost, M. 2003. Biofilm responses to ageing and to a high phosphate load in a bench-scale drinking water system. *Water Research.* 37(6): 1351-1361.

Barbarosa TM and Levy SB 2000. Differential expression of over 60 chromosomal genes in *Escherichia coli* by constitutive expression of MarA. *J. Bacteriol.* 3467-3474.

Baribeau, H., N. Pozos, L. Boulos, G. Crozes, G.A. Gagnon, S. Rutledge, D. Skinner, Z. Hu, R. Hofmann, R.C. Andrews, Z. Alam, S.A. Andrews, R. Dumancic, C. Chauret and E. Warn. 2005. Impact of Changes in Distribution System Water Quality on Disinfection Efficacy. Awwa Research Foundation. Denver, CO.

Baribeau, H., M. Prévost, R. Desjardins, P. Lafrance, and D.J. Gates. 2002. Chlorite and chlorate ion variability in distribution systems. *Jour AWWA*. 94(7): 96-105

Besner, MC, V. Gauthier, P. Servais, and A. Camper. 2002. Explaining the occurrence of coliforms in distribution systems. *Jour AWWA,* 94 (8): 95-110.

bioMerieux, Inc. 2005. API 20E- Indentification system for *Enterobacteriaceae* and other non-fastidious gram-negative rods. *Test Kit Instructions*. REF 20 100/ 20 160.

Boe L, Danielsen M, Knudsen S, Petersen JB, Maymann J and Jensen P.R. 2000. The frequency of mutators in populations of *Escherichia coli. Mutation Research* 448: 47-55.

Box, G.E.P., Hunter, W.G. and Hunter, J.S. 1978. *Statistics for Experimenters: An Introduction to Design, Data Analysis, and Model Building*. John Wiley and Sons.

Brown EW, LeClerc JE, Kotewicz ML and Cebula TA 2001. Three R's of bacterial evolution: how replication, repair and recombination frame the origin of species. *Environ. Mol. Mutagen*. 38: 248-260.

Camper, A., 1995. Factors Influencing Biofilm Growth in Drinking Water Distribution Systems. Ph.D. diss., Montana State University-Bozeman, Bozeman, MT.

Camper, A and W. Jones. 2000. Factors affecting microbial growth in model distribution systems. AwwaRF and AWWA, Denver, CO.

Camper, A. 1996. Factors limiting microbial growth in model distribution systems: laboratory and pilot scale experiments. AwwaRF and AWWA, Denver, CO.

Camper, A. K., P. Butterfield, B. Ellis, and C. Abernathy. 1999. Coliforms in biofilms: an overview. *AWWA 1999 WQTC Proceedings.*

Camper, A.K., Warnecke, M., Jones, W.L. and G.A. McFeters. 1998. *Pathogens in Model Distribution System Biofilms*. AwwaRF, Denver CO

Camper, A., G.A. McFeters, W.C. Characklis and W.L. Jones. 1991. Growth kinetics of coliform bacteria under conditions relevant to drinking water distribution systems. *Applied and Environmental Microbiology*, 57(8):2233.

CDC. 1999. Outbreak of Escherichia coli)157:H7 and Campylobacter amoung attendees of the Washington county Fair-New York, 1999. *MMWR* 48:803..

Craun, GY & Calderon R.L., 2001. Waterborne disease outbreaks caused by distribution system deficiencies. *Jour AWWA*, 93(9):64.

Chauret, C., Volk, C., Creason, R., Jarosh, J., Robinson J. & Warnes, C. 2001. Detection of *Aeromonas hydophila* in a drinking water distribution system: A field and pilot study. *Can. J. Microbiol.*, 47, 782-786

Christensen BB, Sternberg C, Andersen JB, Eberl L, Mooler S, Givskov M and Molin S. 1998. Establishment of genetic traits in a microbial biofilm community. *Appl. Environ. Microbiol.* 64:2247-2255.

Clement, J., Camper A.K. and A. Sandvig. 2002. Influence of distribution system infrastructure on bacterial regrowth. *Proceedings, 2002 AWWA Water Qual. Technol. Conf.* Seattle, WA.

Dykstra, T.S., O'Leary, K.C., Chauret, C., Andrews, R.C., and Gagnon, G.A. 2006. Impact of UV disinfection on biological stability in distribution systems. *J. Env. Engrg & Sci.* In press.

Dykstra, T.S. 2002. Impact of ultraviolet disinfection on the biological stability in drinking water distribution systems. *MASc. Thesis. Dalhousie University, Halifax, Canada.*

Dykstra, T.S., C. Chauret, and G.A. Gagnon. 2002. Hydraulic calibration and fluence determination of model ultraviolet distribution system. *Journal of Environmental Engineering*, 128(11):1046.

Dykstra, T.S., K.C. O'Leary, C. Chauret, R.C. Andrews, and G.A. Gagnon. Impact of UV disinfection on biological stability in distribution systsmes. *J. Env Engrg. & Sci.* In press.

Falkinham, J.O., C.D. Norton, and M.W. LeChevallier. 2001. Factors influencing numbers of mycobacterium avium, mycobacterium intracellulare and other mycobacteria in drinking water distribution systems. *Appl. Environ. Microbiol.* 67: 1225-1231

Fleming, K. and G.W. Harrington. 2003. Using biocide to food ratios as an evaluation and management tool for nitrification control. *Proceedings, 2003 AWWA Water Qual. Technol. Conf.* Philadelphia, PA.

Friedberg, E. C., G. C. Walker, and W. Siede, 1995. DNA repair and mutagenesis. *ASM Press*, Washington, D.C.

Funchain P, Yeung A, Stewart J, Clendenin WM and Miller JH 2001. Amplification of mutator cells in a population as a result of horizontal transfer. *J. Bacteriol.* 183:3737-3741.

Gagnon, G.A. and P.M. Huck. 2001. Removal of easily biodegradable organic compounds by drinking water biofilms: analysis of kinetics and mass transfer. *Water Res.* 35: 2554-2564.

Gagnon, G.A., Rand, J.L., O'Leary, K.C., Rygel. A.C., Chauret, C. and Andrews, R.C. 2005. Disinfectant efficacy of chlorite and chlorine dioxide in drinking water biofilms. *Water Research.* 39:1809-1817.

Gagnon, G.A., and R.M. Slawson. 1999. An efficient biofilm removal method for bacterial cells exposed to drinking water. *J. Microbiol. Methods* 34: 203-214.

Gagnon, G.A., T.S. Dykstra, R.C. Andrews, C. Chauret, and C. Volk 2004. *Impact of UV Disinfection on Biological Stability*. Awwa Research Foundation, Denver CO.

Geldreich, E. E, K.R. Fox, J.A. Goodrich, E. W. Rice, R. M. Clark and D. L. Swerdlow. 1992. Searching for a water supply connection in the Cabool, Missouri, and disease outbreak of Escherichia coli O157:H7. *Water Research*, 26(8):1127.

Giese, N. 2002. *Impact of Ultraviolet Disinfection on Biofilm Growth in a Model Distribution System*. Ph.D. Dissertation, Dept. Civil & Envir. Engrg, University of California, Davis, CA.

Gilbert P and McBain AJ 2003. Potential impact of increased use of biocides in consumer products on prevalence of antibiotic resistance. *Clin. Microbiol. Rev.* 16:189-208.

Hancock REW and Brinkman FSL 2002. Function of *Pseudomonas* porins in uptake and efflux. *Ann. Rev. Microbiol.* 56: 17-38

Harris RS, Feng G, Ross KJ, Sidhu R, Thulin C, Longerich S, Szigety SK, Winkler ME and Rosenberg SM 1997. Mismatch repair protein MutL becomes limiting during stationary-phase mutation. *Genes and Development* 11: 2426-2437.

Hausner M and Wuertz S 1999. High rates of conjugation in bacterial biofilms as determined by quantitative *in situ* analysis. *Appl. Environ. Microbiol.* 65: 1-9.

Helling RB, Janes BK, Kimball H, Tran T, Bundesmann M, Check P, Phelan D, and Miller C. 2002. Toxic waste disposal in *Escherichia coli*. *J Bacteriol.* A184(13):3699-703.

Hengge-Aronis R 2002. Signal transduction and regulatory mechanisms involved in the control of the sigma (S) (RpoS) subunit of RNA polymerase. *Microbiol. Mol. Biol . Rev.* 66:373-395.

Hofmann, R., R.C. Andrews, and Q. Ye. 1999. Impact of *Giardia* inactivation requirements on ClO_2 by-products. *Environ. Technol.* 20, 147-158.

Huck, P.M. and G.A. Gagnon. 2004. Understanding the distribution system as a bioreactor: a framework for managing HPC levels. *International J. of Food Microbiology*. 92(3): 347-353.

Hunter, P.R., J. M. Colford, M. W. LeChevallier, S. Binder and P. S. Berger. 2001. Panel on Waterborne Diseases. *Emerging Infectious Diseases*, 7(3). Retrieved on January 12, 2006 from: http://www.cdc.gov/ncidod/eid/vol7no3_supp/hunter.htm

Hrudey, S.E., P.M. Huck, P. Payment, R.W. Gillham, and E. J Hrudey. 2002. Walkerton: Lessons learned in comparison with waterborne outbreaks in the developed world. *Journal of Environmental Engineering and Science*, 1(6): 397.

HRWC 2005. Halifax Regional Water Commission 2005 Annual Water Quality Report. Halifax, NS.

Jackson DW, Suzuki K, Oakford L, Simecka JW, Hart ME, and Romeo T. 2002. Biofilm formation and dispersal under the influence of the global regulator CsrA of Escherichia coli. *J Bacteriol.* 184(1):290-301.

Kirmeyer, G.,M. Friedman, K. Martel, D. Howie, M. LeChavaller, M. Abbaszadegan, M. Karim, J. Funk, and J. Harbour. 2001. *Pathogen Intrusion into the Distribution System*. AwwaRF. Denver, CO.

Koivunen, J. and Heinonen-Tanski, H. 2005. Inactivation of enteric microorganisms with chemical disinfectants, UV irradiation, and combined chemical/UV treatments. *Water Research.* 39: 1519-1526.

Karim, M.R., M. Abbaszadegan, and M.W. LeChevallier. 2003. Potential for pathogen intrusion during pressure transients. *Jour AWWA*, 95(5):134.

Kotewicz ML, Brown EW, LeClerc JE and Cebula TA 2003. Genomic variability among enteric pathogens: the case of the *mutS-rpoS* intergenic region. *Trends Microbiol*; 11: 2-6.

LeChevallier, M.W. 1990. Coliform regrowth in drinking water: a review. *Jour AWWA* 82(11): 74-86.

LeChevallier, M.W., C.D. Lowry, and R. Lee. 1990. Disinfecting biofilms in a model distribution system. *Jour. AWWA* 82(7): 87-99.
LeChevallier, M. W., N.J. Welch, and D.B. Smith. 1996. Full-scale studies of factors related to coliform regrowth in drinking water. *Journal of Applied and Environmental Microbiology*, 62: 2201-2211.
LeChevallier, M.W. and G.A. McFeters. 1985. Interactions between heterotrophic plate count bacteria and coliform organisms. *Applied and Environmental Microbiology,* 49(5):1338.
LeChevallier, M. W. 1990. Coliform regrowth in drinking water: A Review. 1990. *Jour AWWA,* 83(11): 74-86.
LeChevallier, M.W., C.D. Lowry, and R. G. Lee. 1990. Disinfecting Biofilms in a model distribution system. Jour. AWWA, 83(7):87-99.
LeChevallier, M. W., M. R. Karim, J.E. Funk, and M. Friedman. 1999. Pathogen intrusion into potable water. *AWWA 1999 WQTC Proc*
LeChevallier, M.W., Cameron, S.C. and G.A. McFeters. 1983. New Medium for Improved Recovery of Coliform Bacteria from Drinking Water. *Applied and Environmental Microbiology* 45(2):484-492.
LeClerc JE, Li B, Payne WL and Cebula TA 1996. High mutation frequencies among Escherichia coli and Salmonella pathogens. *Science* 274: 1208-1211.
Levy DD and Cebula TA 2001. Fidelity of replication of repetitive DNA in *mutS* and repair proficient *E. coli. Mutation Research* 474:1-14.
Li B, Ho-Ching TT, LeClerc JE, Dey M, Winkler ME and Cebula TA 2003. Molecular analysis of *mutS* expression and mutation in natural isolates of pathogenic *Escherichia coli. Microbiology* 149: 1323-1331.
Licence, K., K. R. Oates, B. A. Synge and T.M.S. Reid. 2001. An outbreak of *E. coli* O157 infection with evidence of spread from animals to man through contamination of a private water supply. *Epidemiology and infection*, 126:135-138.
Lisle, J.T., S.C. Broadaway, A.M. Prescott, B.H. Pyle, C. Fricker and G.A. McFeters.
1998. Effects of starvation on physiological activity and chlorine disinfection resistance in *Escherichia coli* O157:H7. *Applied and Environmental Microbiology,* 64(12): 4658.
Mack, J. and Bolton, J.R. 1999. Photochemistry of Nitrite and Nitrate in Aqueous Solution: A Review. *J.Photochem.Photobiol.A.Chem*. 128:1-13.
Maenhaut-Michel G, and Caillet-Fauquet P.1982. Effects of Escherichia coli mutator genes mutH, mutL and mutS on 2-aminopurine induced DNA repair.*Biochimie*, 64:739-42.
Martin RG, Jair KW, Wolf RE Jr, Rosner JL. 1996. Autoactivation of the marRAB multiple antibiotic resistance operon by the MarA transcriptional activator in *Escherichia coli. J Bacteriol.* 178(8):2216-23.
Martiny, A.C., T.M. Jorgensen, H.J. Albrechtsen, E. Arvin, and S. Molin 2003. Long-term succession of structure and diversity of a biofilm formed in a model drinking water distribution system. *Appl. Enviro. Microbio.* 69: 6899-6907.
Martel, K.D., Kirmeyer, G.J., Murphy, B.M., Noran, P.F., Kirby, L., Lund, T.W., Anderson, J.L., Medhurst, R. and Caprara, M. 2002. Preventing water quality deterioration in finished water storage facilities. *Journal of the American Water Works Association*. 94(4): 139.
Mathews, C.K., K.E. van Holde, and K.G. Ahern. 1999. Biochemistry 3rd Edition. Addison Wesley Longman, Inc. Don Mills, Ontario.

Matic I, Radman M, Taddei F, Picard B, Doit C, Bingen E, Denamur E, and Elion J 1997. Highly variable mutation rates in commensal and pathogenic Escherichia coli. *Science* 277(5333):1833-4

McFeters, G.A., S.C. Broadaway, B.H. Pyle, M. Pickett and Y. Egozy. 1995. Comparative performance of Colisure™ and accepted methods in the detection of chlorine-injured total coliforms and *E. coli*. *Water Science and Technology*, 31(5-6):259-261.

McFeters, G.A., B.H. Pyle, S. J. Gillis, C.J. Acomb and D. Ferrazza. 1993. Chlorine injury and the comparative performance of Colisure™, Colilert™ and Coliquik™ for the enumeration of coliform bacteria and *E. coli* in drinking water. *Water Science and Technology*,27(3-4):261-265

McFeters, G.A., J. S. Kippin, and M. W. LeChevallier. 1986a. Injured coliforms in drinking water. *Applied and Environmental Microbiology*, 51(1):1-5

McFeters, G.A., M.W. LeChevallier, A. Singh and J. S. Kippin. 1986b. Health significance and occurrence of injured bacteria in drinking water. *Water Science and Technology*, 18(10):227-231..

McFeters, G.A., S.C. Cameron, and M.W. LeChevallier. 1982. Influence of diluents, media, and membrane filters on detection of injured waterborne coliform bacteria. *Applied and Environmental Microbiology*, 43:97-103.

Miller JH, Suthar A, Tai, J,m Yeung A, Truong C and Stewart JL. 1999. Direct selection for mutators in *E. coli*. *J Bacteriol*. 181:1576-1584.

McGuire, M.J., Pearthree, M.S., Blute, N.K., Arnold, K.F. and Hoogerwerf, T. 2006. Nitrification control by chlorite ion at pilot scale. *Journal of the American Water Works Association*. 98(1): 95-105.

Muyzer, G., De Waal, E., and Uitterlinden, A.G. 1993. Profiling of complex microbial populations by denaturing gradient gel electrophoesis analysis of polymerase chain reaction-amplified genes coding for 16S rRNA. *Appl Environ Microbiol*. 59:695-700

Muyzer, G., and Smalla, K. 1998. Application of denaturing gradient gel electrophoresis (DGGE) and temperature gradient gel electrophoresis (TGGE) in microbial ecology. Antonie van Leeuwenhoek. 73:127-141

Muyzer, G., Brinkhoff, T., Nübel, U., Santegoeds, C, Schäfer, H, and Wawer, C. 1998. Denaturing gradient gel electrophoresis (DGGE) in microbial ecology. Molecular Microbial Ecology Manual. Kluwer Academic Publishers, Netherlands. 3.4.4:1-27

Muyzer G 1999. DGGE/TGGE a method for identifying genes from natural ecosystems. *Curr Opin Microbiol*. 2(3):317-22.

Nagy, L.A. and B.H. Olson. 1985. Occurrence and significance of bacteria, fungi and yeasts associated with distribution pipe surfaces. *Proceedings, AWWA Water Quality Technology Conference*, American Water Works Association, Denver, CO.

Norman, T.S., Harms, L.L. and Looyenga, R.W. 1980. The use of chloramines to prevent trihalomethane formation, *J. Am. Water Works Assoc.* 72(3):176–180.

Norton, C.D. and M.W. LeChevallier. 1997. Chloramination: its effect on distribution system water quality. *Jour AWWA* 89(7): 66-77.

Notley-McRobb L, Pinto R, Seeto S, and Ferenci, T. 2002. Regulation of *mutY* and nature of mutator mutations in *Escherichia coli* populations under nutrient limitation. *J. Bacteriol.* 184: 739-745.

Nowell, LH and Hoigne, J 1992. Photolysis of aqueous chlorine at sunlight and ultraviolet wavelengths. II. Hydroxyl radical production. Water Research 26 (5): 599-605.

Olstadt, J.M., J. Standridge, S. M. Kluender and J. Schauer. 2005. A comparison of USEPA-approved enzyme-based total coliform and *E. coli* tests for groundwater monitoring. *AWWA 2005 WQTC Proc.*

Ormeci, B., Ducoste, J. and Linden, K. 2005. UV disinfection of chlorinated water: impact on chlorine concentration and UV dose delivery. *Journal of Water Supply: Research and Technology – AQUA.* 54(3): 189-199.

Parkinson, A., Roddick, F.A. and M.D. Hobday. 2003. UV photooxidation of NOM: issues related to drinking water treatment. *Journal of Water Supply: Research and Technology – AQUA* 52(8):577-586.

Parsek, M.R. and C. Fuqua. 2004. Meeting Review- Biofilms 2003: Emerging themes and challenges in studies of surface- associated microbial life. *Journal of Bacteriology,* 186(14):4427.

Pintar, K.D. and R.M. Slawson. 2003. Effect of temperature and disinfection strategies on ammonia-oxidizing bacteria in a bench-scale drinking water distribution system. *Water Research,* 37(8):1805-1817.

Potenski CJ, Gandhi M, Matthews KR 2003. Exposure of *Salmonella enteritidis* to chlorine or food preservatives increases susceptibility to antibiotics. *FEMS Microbiol Lett.* 220(2):181-6.

Pozos, N., K. Scow, S. Wuertz, and J. Darby. 2004. UV disinfection in a model distribution system: biofilm growth and microbial community. *Water Research,* 38: 3083-3091.

Pryor, M., Springthorpe, S., Riffard, S., Brooks, T., Huo, Y., Davis, G., Sattar, S.A. 2004. Investigation of opportunistics pathogens in municipal drinking water under different supply and treatment regimes. *Water Science Technology.* 50(1)83-90.

Pryor, M, Springthorpe, S, Riffard, S, Brooks, T, Huo, Y, Davis, G and Sattar, S.A. 2003. Investigation of opportunistic pathogens in municipal drinking water under different supply and treatment regimes. *Presented at International Water Assoc. Biofilms Conf.*, Sept 2003, Cape Town, South Africa.- Accepted for publication in Water Science and Technology, 2004.

Rand, J.L., Gagnon, G.A., Cantwell, R., Liere, J., Hofmann, R., Andrews, R. and VanderMarck, M. 2005. Effect of UV treatment on biological fouling in a California aqueduct. AWWA Annual Conference and Exposition. San Francisco, CA.

Ribas F, J. Perramon, A. Terradillos, J. Frias, and F. Lucena. 2000. The Pseudomonas group as an indicator of potential regrowth in water distribution systems. *J Appl Microbiol.* 88(4):704-10.

Schawrtz, T., S. Hoffmann and U. Obst. 2003. Formation of natural biofilms during chlorine dioxide and UV disinfection in a public drinking water distribution system. *J. App. Microbiol.* 95: 591-601.

Schembri MA, Kjaergaard K and Klemm P 2003. Global gene expression in Escherichia coli biofilms. *Molec. Microbiol.* 48: 253-267.

Schwartz, T., S. Hoffmann, and U. Obst. 1998a. Formation and bacterial composition of young, natural biofilms obtained from public bank-filtered drinking water systems. *Water Research* 32: 2787–2797.

Schwartz, T., S. Kalmbach, S. Hoffmann, U. Szewzyk, and U. Obst 1998b. PCR-based detection of mycobacteria in biofilms from a drinking water distribution system. *J. Microbiol. Meth.* 34: 113–123.

Sharp, R.R., A.K. Camper, J.J. Crippen, O.D. Schneider, and S. Leggiero 2001. Evaluation of drinking water biostability using biofilm methods. *ASCE – J. Env. Engrg.* 127(5): 403-411.

Shaw, J.P., J. P. Malley Jr. and S. A. Willoughby. 2000. Effects of UV irradiation on organica matter. *Jour. AWWA*, 92(4): 157.

Shaw, S., and S. Regli. 1999. US regulations on residual disinfection. *Jour AWWA*, 91(1): 75-80.

Singh, A., and G. A. McFeters.1992. Detection methods for waterborne pathogens. In *Environmental Microbiology*. Wiley-Liss, Inc., New York, New York: 125-156.

Skinner, D.M. 2003 Impact of Water Quality Changes in the Distribution System on Disinfectant Efficacy. Masters thesis. Dept. of Civil Engineering, Dalhousie University, Halifax, NS.

Storz, G., and R. Hengge-Aronis. 2000. Microbial response to acid stress. In *Bacterial Stress Responses*. ASM Press, Washington, D.C.: 99-115.

Taddei F, Matic I, Godelle B and Radman M 1997. To be a mutator, or how pathogenic and commensal bacteria can evolve rapidly. *Trends Microbiol.* 5: and Cebula TA and LeClerc JE (1997) Response to above article.

Tanabe K, Kondo T, Onodera Y. and Furusawa M. 1999. A conspicuous adaptability to antibiotics in the Escherichia coli mutator strain, dnaQ49. *FEMS Microbiol. Lett.* 176: 191-196.

Trussell R.R. 1999. Safeguarding distribution system intergrity. *Jour AWWA,* 91(1):46.

Standard Methods for the Examination of Water and Wastewater 21st Edition. 2005. APHA, Washington, D.C., USA.

USEPA. 2002a. Health risks from microbial growth and biofilms in drinking water distribution systems. Retrieved January 4, 2006 from: http://www.epa.gov/safewater/tcr/pdf/biofilms.pdf

USEPA. 2002b. Potential contamination due to cross-connections and backflow and the associated health risks. Retrieved January 3, 2006 from: http://www.epa.gov/safewater/tcr/pdf/ccrwhite.pdf

World Health Organization (WHO) 2004.

Tsui H-CT, Feng G and Winkler ME 1997. Negative regulation of *mutS* and *mutH* repair gene expression by the Hfq and RpoS global regulators of Escherichia coli K-12. *J. Bacteriol.* 179.7476-7487.

Wilczak, A., J.G. Jacangelo, J.P. Marcinko, L.H. Odell, G.J. Kirmeyer, and R.L. Wolfe. 1996. Occurrence of nitrification in chloraminated distribution systems. *Jour AWWA* 88(7): 74-85

Volk, C. J. and M. W. LeChevallier. 2000. Assessing biodegradable organic matter. *Jour AWWA,* 92(5):64.

Volk, C. J., R. Hoffman, C. Chauret, G.A. Gagnon, G. Rangers and R. C. Andrews. 2002. Implementation of chlorine dioxide: effects of the treatment change on drinking water quality in a full-scale distribution system. *Journal of Environmental Engineering and Science*, 1(5):323.

Williams, M.M. and E.B. Braun-Howland. 2003. Growth of Escherichia coli in model distribution systems biofilms exposed to hypochlorous acid or monochloramine. *Applied and Environmental Microbiology*, 69(9): 5463.

Werdehoff, K.S., and Singer, P.C. 1987. Chlorine dioxide effects on THMFP, TOXFP and the formation of inorganic by-products. *Journal of the American Water Works Association.* 79(9): 107-113.

Zimmer J.L. and Slawson R.M. 2002. Potential repair of *Escherichia coli* DNA following exposure to UV radiation from both medium- and low-pressure UV sources used in drinking water treatment. *Appl Environ Microbiol.* 68(7):3293-9.

ABBREVIATIONS

ANOVA	analysis of variance
AOB	ammonia oxidizing bacteia
AR	annular reactor
Awwa	American Water Works Association
AwwaRF	Awwa Research Foundation
BAC	biologically active carbon filter
BOM	biodegradable organic matter
°C	degrees Celsius
cfu	colony forming unit
CI	cast iron
Cl_2	chlorine
ClO_2	chlorine dioxide
cm^2	centimeter squared
DBP	disinfection by-products
DGGE	denaturing gradients gel electrophoresis
DNA	deoxyribonucleic acid
DOC	dissolved organic carbon
E. coli	*Escherichia coli*
GAC	granular activated carbon
h	hours
HPC	heterotrophic plate count bacteria
HRT	hydraulic retention time
HRWC	Halifax Regional Water Commission
H_2O	water
in	inches
J	joules
K12	*Escherichia coli* K12
L	liters
m	meter
mm	milimeter
m-FC	fecal coliform membrane filtration method
mg	miligrams

min	minutes
mm	millimeters
MMR	mismatch repair mutant
mL	millilters
MPN	most probable number
µm	micrometers
µm	micromolar
µL	microliters
$NHCl_2$	monochloramine
N/m^2	Newtons per meter squared (sheer stress)
nm	nanometers
NOM	natural organic matter
NTU	nephelometric turbidity units
PC	polycarbonate
PCR	polymerase chain reaction
PBS	phosphate buffer solution
PVC	polyvinyl chloride
RDP	Blast and Ribosomal Database Project II
rpm	revolutions per minute
TOC	total organic carbon
TTHM	total trihalomethanes
USEPA	United States Environmental Protection Agency
UV	ultraviolet
w/v	weight per volume

IWA Publishing's authorised EU representative for General Product Safety Regulations is Diane D'Arras, 15 rue Duret, 75116 Paris, France, e-mail: safety@iwap.co.uk.

Printed and bound by CPI Group (UK) Ltd, Croydon, CR0 4YY
17/09/2026
02218808-0018